THE KITCHEN

THE KITCHEN

A journey through history
in search of the perfect design

John Ota

appetite
by RANDOM HOUSE

Library and Archives of Canada Cataloguing in Publication is available upon request.

ISBN: 9780525609896
eBook ISBN: 9780525609919

Photo credits: page 121: Tenement Museum; page 184: *Georgie O'Keeffe's Abiquiu House, Kitchen Interior*, 2007. Herbert Lotz. © Georgia O'Keeffe Museum; page 246: Louis Armstrong House Museum and Archives. All other photos © John Ota.

Book design by Leah Springate

Printed and bound in Canada

Published in Canada by Appetite by Random House®,
a division of Penguin Random House Canada Limited

www.penguinrandomhouse.ca

10 9 8 7 6 5 4 3 2 1

appetite
by RANDOM HOUSE

Penguin
Random House
Canada

For Franny

kitch-en\ 1: a place (such as a room) with cooking facilities

—*Merriam-Webster Dictionary*

It is certainly the most loved and most used room in the house.

—JULIA CHILD

CONTENTS

INTRODUCTION

MY LOVE FOR KITCHENS began at an early age. My happiest childhood memories are of spending time in the kitchen in our family home. I grew up in a working-class neighbourhood in Toronto. My dad built boats and my mom was a high school art teacher. We were not wealthy, but I never felt poor, and the shelves of the refrigerator always groaned from the weight of stews, knishes, dai baos, sushi, coq au vin and lasagnas. We loved to eat. My two younger brothers and I ate so continuously that my dad used to watch in horror as his paycheque was devoured before his eyes.

The kitchen was the hub of the house. All breakfasts, lunches, dinners and snacks were consumed there. But almost more important than the meals, it was also the place of family discussions, talks about holiday plans, hockey games, report cards, politics, hosting neighbours and relatives and my mom's daily report about her teaching day.

Our kitchen was the opposite of today's sleek white ideal. It was semi-organized chaos surrounded by loud orange-and-green wallpaper. Flour, sugar, tea and coffee in scratched-up canisters held together with masking tape out on counters; salt, pepper, sugar in open plastic cups by the stove; bottles of olive oil and vinegar everywhere; knives, can openers, bottle openers, rice cooker within easy reach; mountains of cookbooks stacked against the walls. It was a kitchen for cooking in, not for being photographed.

From my architecture and historic preservation studies, and later at work, I found that most people did not share my reverence for the kitchen. Most would ooh and aah over the elaborate architectural details elsewhere in historic houses. I enjoyed those too, but I was always more drawn to the room others ignored. When I told colleagues about my obsession, they would tilt their heads and look at me strangely, as if to say, "You're not a real preservationist." To most of them, historical preservation was about the entrance halls, palm rooms and parlours.

I've worked in the field of architecture for over forty years, as a writer, designer, historic preservationist and curator, so people often want to talk to me about their house renovations. The first question I ask them is, "What is your perfect kitchen?" It's the ice breaker. The kitchen is the subject of everyone's strongest, truest opinions. Without exception, when asked about the kitchen, people express their wants, visions and wish lists to me.

"I'd love a long, sweeping island."

"I need more counter space."

"A big stainless steel refrigerator would be wonderful."

Their real passion is not for the living room or the bathroom or the dining room. Renovations begin in and flow from the kitchen, because it is the centre of everything that happens in the house.

Each of us has a vision of the perfect kitchen. I have designed, written about and fallen in love with small and large kitchens, old and new kitchens, with big budgets and small, for clients and for friends. And I have met people with unique ideas for the perfect kitchen.

In one project, all the building materials, appliances and floor plans were considered with the family dog in mind. I understood. I was a dog in my previous life. On another high-end project, the client specified mirrored tile for the backsplash. When I commented that the mirrored tile might be a problem to keep clean during cooking, she shrugged and replied, "So who cooks?"

Well, I do. Cooking—and eating—are among my greatest pleasures.

I make no great claims for myself as a cook. I mostly rustle up straightforward roasts, soups and salads. But I also love experimenting in unfamiliar (to me) culinary areas (currently, Japanese delicacies like grilled sea bream with miso). My favourite place to be is the kitchen. To me, the kitchen is not just the most important room in my house. It is the centre of my life.

In 2005, my wife, Franny, and I built our own house in a contemporary style. At the time, to control costs, we held back on some nice things for the kitchen. But today our lives have changed. Most evenings we cook at home, our entertaining has happily increased, and although I do most of the cooking, Franny has been doing more.

With both of us spending more time around food, Franny wants to renovate the kitchen. It's a cramped and crowded space for two people. And while I've grown accustomed to its inadequacies, Franny views the room with the critical eye of the newly converted.

It's awkwardly laid out, she says. Its constituent parts are "not in the right place."

Let's face it—my wife *hates* the kitchen.

I find this disturbing, because I try to keep my wife happy.

And so I embarked on a quest. I decided to explore examples of excellent kitchen designs from throughout North American history in order to learn from them so I could improve our own. I would delve into the origins of the kitchen and examine how its architecture evolved in response to new appliances, cooking methods and the shift from cooking being done by invisible servants in separate buildings to its being the one activity that draws the family together.

And because I wanted to do more than just put together different elements from great kitchens, I began my quest by distilling a raison d'être.

I wrote down the three things I wanted my perfect kitchen to do:

Stimulate creativity: Sriracha, native wild rice, dandelion greens—every time I turn around I want to explore another new ingredient. I need updated kitchen equipment, like a bigger kitchen fan to draw away smoke so that I can blacken catfish. We would experiment more with elaborate dishes like croquembouche if we had laid out our counter more efficiently. I'd love for the kitchen to stoke our cooking passion.

Make cooking easier: When we entertain guests we do our best to make sure they have fun. But if the cooking is easier while we're in the kitchen, we'll have more fun too.

Encourage celebration: I would love a kitchen that brings people together—laugh, eat, cook, share, connect with memories, feelings and our past. My mom had a cluttered kitchen and more than good food came out of it. I want a kitchen floor we can dance on.

This book and the kitchen it will inspire are gifts to my wife. But this book is also for everyone who loves cooking and eating and wants a better understanding of their kitchen. Or, indeed, wants a better kitchen.

Everybody has something that turns their crank, and while I am honoured to have worked on a variety of building types in my life, including sports domes, museums, art galleries, churches, courthouses and the tiki bar of a pool cabana, my heart has always been in the kitchen.

I especially like historic kitchens. Sometimes, as I walk around them looking at the appliances, the implements, the knick-knacks, I can almost hear the past occupants speaking to me down the centuries.

I love talking to people about their kitchens—the type of sink they chose, the stove, the fridge, the flooring, how they entertain. Most of all, I love talking to people about what they make to eat in there.

Oh, well. I can't help it. I love kitchens. And I love my wife.

I went in search of our perfect kitchen.

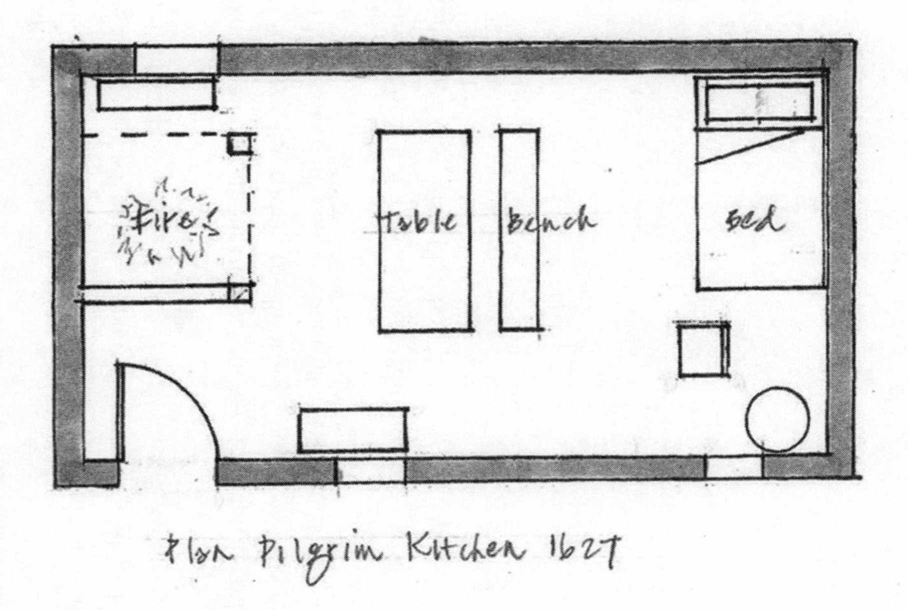

Plan Pilgrim Kitchen 1627

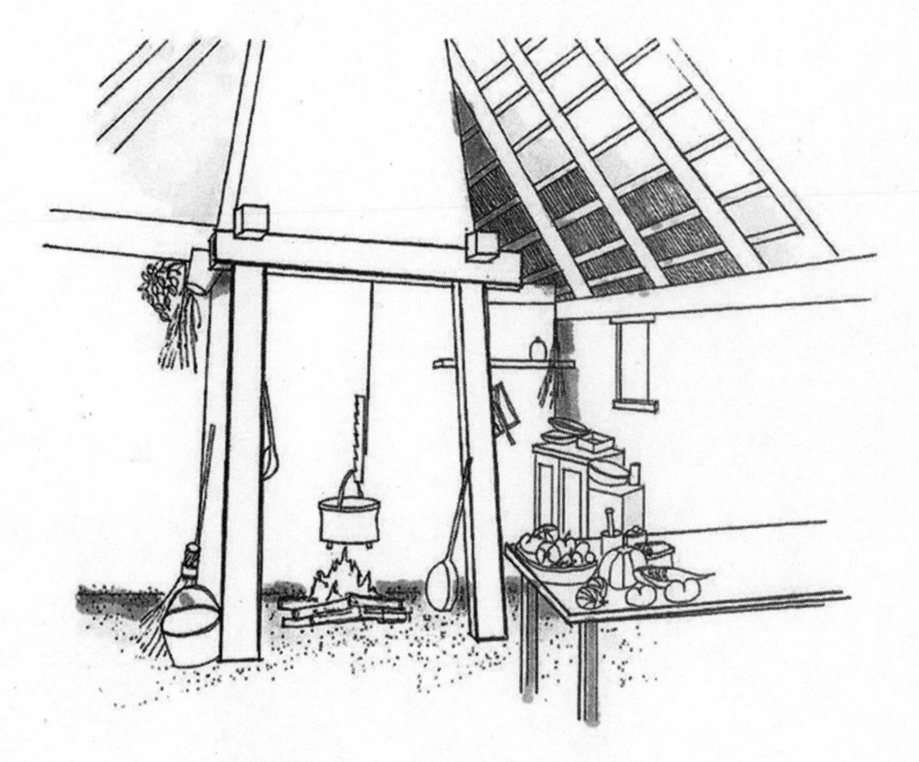

Interior Pilgrim Kitchen 1627

1

PILGRIM KITCHEN, 1627

Plymouth, Massachusetts

I AM STANDING IN PLYMOUTH, Massachusetts, looking out at the Atlantic coastline where the Pilgrims rowed ashore in 1620. I am breathing in the same salty air, feeling the same wind on my face, and I am filled with a sense of connection with the past. As I scan the horizon and walk the dirt paths, I am struck by amazement at the deeds and actions of both the Pilgrims and the people of the Wampanoag tribes.

Even though there were earlier settlements in the New World, Plymouth for me is the spiritual beginning of the United States. I have wanted to visit this place for a very long time.

There could be no longer shot for survival than that faced by the Pilgrims in 1620. The odds against them were so high, I doubt Vegas would even have posted them. Amid turmoil, setbacks and confusion, the Pilgrims set out from England to make their famous voyage across the Atlantic Ocean aboard the *Mayflower* and founded Plymouth Colony. I think it's one of the most fascinating stories of humankind.

And this is why I love visiting historic sites. My daily routine is thousands of miles away from exploring new worlds, building new settlements and protecting winter stores with my life.

I am starting my quest for the perfect kitchen at the beginning—a 1627 Pilgrim house at Plimoth Plantation, a museum of a seventeenth-century Pilgrim colony at the base of Cape Cod. I want to see an early American kitchen in the place where it all began. I need to understand how the Pilgrims lived, what they ate, how they prepared their meals. And to try once again to imagine what courage and determination it took for them to risk coming here in the first place.

My face warmed by the sun rising over the Atlantic, I sit sipping my coffee on a bench outside the entrance to Plimoth Plantation. Although the town's name is now Plymouth, there were no firm rules for English spelling in the early 1600s. "Plimoth" was used by the colony's second governor, William Bradford, and is the version adopted by the museum.

It is seven thirty in the morning and the plantation doesn't open until nine. I am the only person at the visitor centre's front doors. A caretaker walks by and gives me a friendly smile. I'm sure he thinks I'm nuts to be here so early.

Primed to cook Pilgrim dishes in a 1627 kitchen, I have arranged to meet Kathleen Wall at the Plimoth Plantation. I know from her bio that she is a culinary historian, an expert in food and cooking techniques of the *Mayflower* settlers, and I have signed up for her Hardcore Hearth Cooking class.

"Hardcore Hearth Cooking"—that's what I want. I want to see, smell, taste, hear and eat the Pilgrim kitchen experience, close-up, with no compromises and no modern interventions.

Finally the doors open and I see coming toward me a young woman with braided hair. She is wearing an apron and carrying cooking pots and bags of vegetables. This can only be my Pilgrim cooking instructor. Kathleen is instantly likable, and we shake hands with big smiles. I tell her that I am really pumped to do this cooking class.

Kathleen suggests that I walk up the hill to the Pilgrim village, acclimatize myself to the 1620s and then join her inside the house closest to the settlement's entrance. In the meantime, she says, she will build a fire and set up the kitchen inside that house for our cooking class. As we part, I can feel the caffeine from my morning coffee kick in. I am elated to be at Plimoth.

Kathleen walks away, but then stops and turns and warns me with a big smile, "We'll be cooking in true 1627 fashion. We'll be doing a lot of manual labour in the same way that the Pilgrims worked to prepare their food. It'll take a fair amount of time."

I smile back at her and nod. "That's OK. I'm here to cook like a Pilgrim."

I set out toward the Plimoth Plantation village along a dirt path that meanders through an oak forest. As I walk, I try to get into a historical frame of mind. I know the Pilgrims sailed from England in 1620, when ships from England, France, Spain and Holland were sailing the high seas, jockeying to establish settlements in the New World. The Wampanoag people had lived in southeastern Massachusetts for more than twelve thousand years in seventy villages around Cape Cod. Above all, in 1620, the United States of America was not even a twinkle in anyone's eye.

In the distance, I see a palisade of wood poles rising out of a hillside, and the path leads me

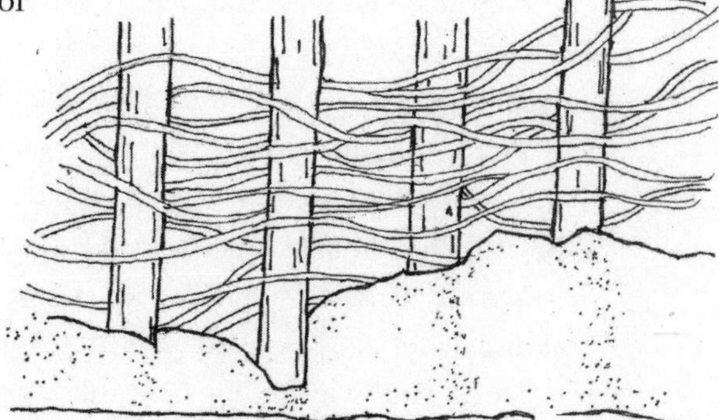

through the gates of a fort. In an instant, I find myself in front of a panoramic view of wood cottages spilling down a main street to the waters of Cape Cod Bay.

I walk down the street toward the ocean. On both sides of me along the dirt road, I count eighteen wood structures, including houses, outbuildings and a large meeting house. They are stark boxes topped by peaked roofs that rise sharply into the sky. Thatched roofs and weathered wood walls give the houses a monochromatic grey wash.

In the background, cattle and sheep roam in the fields. There's a strong scent of woodsmoke in the air. I can hear the clucking of chickens and the hammering of nails into planks. The scene matches what I imagine a Medieval English village looked like.

The Pilgrims called their town a plantation, and farming was a major part of their lives. Each house has its own garden plot of herbs, greens and root vegetables in neat rows. There are fields just outside of town, where corn, wheat and hay are growing. Many of the Pilgrims had come from cities or towns in England and quickly had to learn how to farm.

As I look around, I am surprised by the complete absence of log cabins, which feature in every illustration of Pilgrim life I've ever seen. But it turns out that those log cabins are a myth. The Pilgrims built what they were used to living in, which were English cottages.

Ahead on the road, I see Kathleen bustling around outside a house making final arrangements for our class. She welcomes me to the Pilgrim house where we will be cooking a 1620s meal. While we chat outside, I get a good look at the house's exterior and the Pilgrims' construction methods. I love getting my hands on building materials, foundations, wood details, roofing and windows and figuring out why people built their houses the way they did.

When the settlers arrived in Plymouth in late December, they set about building houses with the axes, saws and hammers they had brought with them. They cut down trees in the forests and made cottages with a strong oak frame, chimney and steep roof rafters. I cannot imagine how they built their houses by hand—in the middle of winter.

To protect the exterior walls, the Pilgrims made clapboards and nailed them to the frame of the house. Roofs were thatched with bulrush to keep out the sun, wind and rain. Construction of each house, thatch and all, took two to three months. The chimney of the house I'm looking at is clad in clapboard like the rest of the building. It's located above the front gable and tells me where the fireplace is inside.

The Pilgrims built houses that have no hint of embellishment or individual expression. There is no welcome ornament at the front door or brackets at the roof overhang. The simple door of vertical boards fastened to horizontal battens presents an attitude of self-effacement and humility. This is an architecture that reflects a Puritan culture of work and devotion, the central themes of Pilgrim life. Hunting, milling, leather work, carpentry, cutting wood, clearing pastures, shearing sheep—there was always work to do. Decorating a house was considered a waste of time in God's eyes.

Kathleen guides me to the front door and my excitement builds as I am about to enter my first Pilgrim house. But as I step inside the doorway and try to look around, I find I am in complete darkness.

Eventually my eyes adjust and I am able to peer into a single room of about fourteen feet by twenty feet. I can see that the interior is rough and spare. Slivers of sunlight peek through a few small windows that are shuttered with oiled paper.

But as I gaze up into the fifteen-foot-high rafters, I am elated to be surrounded by the pots, pans, utensils and gritty atmosphere of this early seventeenth-century kitchen. I am standing in a Pilgrim kitchen at last. Or at least, I am in a house that contains just one room. I can see from the wood bed at the back that as well as a kitchen, this room was used as a bedroom, a living room and a dining room, for a family of perhaps six to ten people.

Suddenly I start to cough and sneeze and my eyes tear up. Black smoke is everywhere and I am having trouble breathing. I turn around to see an open fire burning in a pit against the wall at the front of the house. I am embarrassed to be coughing in front of Kathleen—embarrassed to be such a city guy, accustomed to pollution, maybe, but not to actual smoke.

The fire was crucial to life in these houses—the visual focal point but also the source for heat to cook food, warmth during the winter months and light at night. Elevated about eight feet above the open fire is a chimney shaft, built of wood and covered in clay, that reaches through the rafters. Kathleen has hung pots above the fire from a single branch embedded in the sides of the chimney.

Suddenly I start to cough and sneeze and my eyes tear up. Black smoke is everywhere and I am having trouble breathing.

There is no brick fireplace. This is simply an indoor campfire, built on a flat rock sitting on the dirt floor that extends into the rest of the house. Open fires burned the houses at times, but the Pilgrims had no bricks and no time to build a fireplace. The main thing was to have shelter while they planted crops that would keep them alive.

A sturdy wood table in the middle of the room acts as the preparation counter. Covered in bowls, small pots of herbs and a mortar and pestle, this is the only work surface. One bench, about four feet long, and a single wood chair are arranged around the table for seating. Scattered on the dirt floor are pots, urns and a pail of water to control the fire.

The walls are framed in wood and have a mud-grey plaster finish. To make the walls, the Pilgrims wove a mesh frame of sticks called wattle inside the house frame. Next they made a mortar mixture of clay, earth, grasses and water called daub. They forced the daub mortar into the wattle mesh to fill the wall and then smoothed the interior finish.

To the right of the fire, a window allows a small amount of sunlight and air into the cottage. Like all the windows, it's minute—a response to the cold weather and the high cost of glass. I am surprised that one small chest under the window is the only place to store plates and mugs.

There are no mysteries in this kitchen, no grandeur. I can see that whatever is in the room is there to support cooking. The aesthetic is purely functional.

Yet I find there is something refreshingly genuine about it. Hand built, solid, crafted with materials from the surrounding environment, it is the original organic kitchen. It has rugged good looks. Unlike some vinyl-covered imitations of today, the overhead beams and rafters are real, and the timbers still show axe marks from when the bark was removed. The very basic fire is immediate and formidable and it reminds us of how real cooking started. The single table in the middle of the kitchen might look old and utilitarian, but it is the predecessor of the contemporary kitchen island.

There is no way I would want to go back to cooking over an open fire, but sometimes I feel that I take too much for granted. Quite often, I am stressed for time when making dinner. In my race for efficiency and convenience, I whiz, blend and beat with a plethora of utensils and appliances that can cause me to become detached from what I am really doing—detached from the satisfaction of cooking. The Pilgrim kitchen at Plimoth puts me back in touch with basic architecture and with basic cooking.

—

As Kathleen arranges ingredients around the central kitchen table, she announces that one of the dishes we will be preparing is duck. My mouth waters instantly. I can practically taste the crispy, chewy duck skin.

But then she tells me that we'll be boiling it.

"What?" I had pictured a sizzling duck roasting on a spit with glistening duck fat dripping into the open fire.

"Colonists boiled everything, including geese and salads," says Kathleen. "They used the cooking liquid for soups and sauces."

"OK. No problem. I'm ready to try anything."

"If you wanted meat on the table, you had to shoot it yourself," says Kathleen. "Waterfowl such as mallard and teal ducks and cormorants were plentiful in Cape Cod Bay, and the Pilgrims hunted them with their rifles." Once the men shot and gathered the ducks, it was the responsibility of the women to pluck and gut.

I am grateful that Kathleen has spared me the plucking and gutting duties by bringing an already cleaned duck for our cooking session.

I adore duck. I still remember the first time I tasted the roast duck that my mother made for a Sunday dinner one cold autumn evening. The meat was dark, tender and full of deep flavour. We marvelled at the fattiness of the crispy skin. My brothers and I fought over the drumsticks.

There are so many things we could do with this gorgeous bird other than boiling it. I begin to imagine the fantastic possibilities. We could make pan-seared duck breasts in berry sauce or canard à l'orange or Peking duck with scallions and pancakes (I love the crunchy skin with hoisin) or roast duck stuffed with herbed rice and chestnuts or duck confit or a pampered sous vide duck breast.

Kathleen smiles at me and directs me to plop the duck into a deep pot and pour water over it until the bird is half submerged. She covers the pot and places it on a vertical hanger above the open flame. The hanger has several hooks at different levels that allow the cook to hang the vessel lower or higher to control the heat. Kathleen tells me that after the liquid comes to a boil we will be adding vegetables to the pot.

Oh well. Farewell, my honourable waterfowl friend. I pray there will be some leftovers. I could make my favourite—duck fried rice with finely chopped green onions sizzled in sesame oil and laced with a partially beaten egg.

A stiff breeze from the side window makes the fire spit and spark. Embers float out. For a moment everything in the house turns black and my eyes smart from the hot air and smoke blowing into my face. Coughing and rubbing my eyes, I wonder how the Pilgrims could have cooked like this every day in what is essentially a smokehouse.

Of course, my wonderful Pilgrim mentor continues cooking like nothing has happened.

I step outside to get some fresh air.

The Puritans were a religious group who left their homeland in the 1600s because they wanted to practise a simpler faith than that of the Church of England. They were named after their desire to "purify" the church. They were not dubbed Pilgrims until the 1820s, on the bicentennial of the founding of the colony.

Mayflower arrived in New England on November 11, 1620, after a difficult voyage that took two months. The ship anchored off what is now Provincetown, and after some exploring of the coast, it dropped anchor at Plymouth Harbor on December 16. A week later, on December 23, parties of men began to go ashore to build houses and returned to the ship each night. Women and children stayed onboard through the first winter. All suffered from the cold weather and weakened health. They named the place New Plimoth, after their last port of call back home.

As they built their village, they met a Native American named Squanto, who was able to communicate with the Pilgrims in English. Squanto had been kidnapped by English explorers in 1614 and learned the language while a captive in London.

Squanto is a hero in the Pilgrim story. He taught them to plant corn, and where to fish and hunt. The Pilgrims would not have survived without him.

During that first winter, many colonists fell ill with scurvy or pneumonia. Of the original 102 passengers, only 52—and only four of them women—survived the first year in Plymouth.

Food was a continuous preoccupation for the colonists. Their daily life revolved around it: hunting, fishing, farming, cooking. Preserving was important too, in order to ensure supplies throughout the year. Salting, smoking and drying were used to preserve meat and fish. Vegetables were pickled in vinegar and sugar.

Small wonder that the kitchen was their dwelling's centre and held the family together.

Kathleen calls me back into the house and tells me the next task is to start preparing our quails.

I take a step backwards when she unsheathes an enormous samurai sword and drops it on the table with a thud. On closer examination, I see it is not a sword at all, but a long steel rod with pointed ends used to skewer quails and rotate them on the fire.

Kathleen unwraps a package of about a dozen quails, small birds that the Pilgrims hunted at Plimoth. Again I am grateful that the birds have been pre-cleaned and plucked, a luxury that the Pilgrims did not enjoy. Kathleen proceeds to stand the rod upright on the table and impale the quails in one long row.

The skewered birds are then covered in globs of butter. Then the rod is placed on two steel brackets over the low flame.

Kathleen bends down to place a long, oblong dish under the birds to catch the drippings, which she will use to baste the birds.

I watch the quails as they begin to roast and drip butter and their own fat beside the low fire. Kathleen tells me that the Pilgrims brought butter with them on the *Mayflower*. "It was a staple in the seventeenth century," she says. "Sailors were allotted one pound of butter or one pound of cheese per day." A high-fat diet on ships, with not a fresh vegetable in sight.

But maize, or corn, she says, became a staple for the Pilgrims, and they baked it into bread and cakes. Squanto taught the Pilgrims the Native method of growing corn alongside beans and pumpkin. After the land was cleared for farming, the Pilgrims fertilized the soil by placing fish in the ground—something else Squanto taught them. Corn seeds were planted in rows and eventually the corn stalks became supports for bean plants. The beans further fertilized the soil and the pumpkin leaves shaded the corn roots. The "Indian corn" that grew was made up of multicoloured kernels of red, yellow and black on the same ear. The corn was dried and then ground into flour.

I am in a culinary trance, watching fat drip off the roasting quails, when Kathleen calls me and leads me outside. "Now it's time to get the vegetables going," she says. In the bright sunlight and fresh air, a wood table is covered in fresh vegetables, including large bulbous turnips, leafy cabbages, oblong squash and enormous orange pumpkins. She harvested these vegetables just minutes ago from the Plimoth gardens; they are the same varieties of vegetables the Pilgrims would have eaten in the 1620s.

We begin the task of cleaning, peeling and chopping. Kathleen hands me a bucket of water, a rag and an old knife, and I start my work as a Pilgrim cook. It's hard manual labour. For one thing, the turnips are covered in dirt. "The Pilgrims kept them in the ground as long as possible to improve their taste,"

Kathleen tells me. I can't remember ever seeing such large and colourful turnips in a supermarket.

My turnip washing goes well, but hollowing the pumpkins and peeling is a different story. First, I have to cut a round hole in the top of the large orange globe and then reach in with my hands to remove the seeds and stringy pulp. Next, I chop up the rock-hard flesh into small segments. The final task is peeling off the pumpkin skin with an old knife. The skins are tough, and my hands get cramped from the intense gripping of the knife against the rock-hard pumpkin surface. The big, awkward-to-hold ones with waxy skin are a particular challenge.

The Pilgrims did not use cutting boards, and I do all my chopping directly on the wood table, pressing the knife down hard on the unforgiving veg. After about fifteen minutes I am grunting and sweating. But Kathleen and I, it turns out, have another forty-five minutes of exertion ahead of us. I dutifully hack away, all the while thinking about how easy it is in present-day life to open a cellophane bag and have the turnip, squash or pumpkin tumble out, already cleaned, peeled and chopped.

At the end of my ordeal, I pat Kathleen on the back as we proudly survey our collection of pumpkin and turnip chunks. I am left with a strong impression of how much effort the Pilgrims put into cooking their meals. Add to this lifting heavy pots of boiling water, fetching wood for the fire, plucking and cleaning chickens, turkeys and geese and inhaling clouds of black smoke over an open flame—cooking was a big job.

And food preparation and cooking would be never-ending, all day from sunrise to sundown. Even though it was the 1620s, the Pilgrims' dining routine of three meals daily was not that different from ours today. On rising, the Pilgrims ate cheese, bread and butter. The most substantial meal of the day took place around noon and included meat or fish with turnips or pumpkin and bread. In the evening they ate a light serving of what remained from the mid-day meal. The Pilgrims ate with their hands. They had spoons and knives, but forks were not popular in North America until the late 1700s. For picking up hot meat and wiping their fingers, they used a cloth draped over the shoulder.

As I stand in my sweat-soaked T-shirt, Kathleen tells me that our next task is to prepare the spices for cooking. Pilgrims brought a surprising number of cooking spices with them, including salt, pepper, cinnamon, nutmeg, mace and ginger.

It has been a novel experience to be a sous chef "alfresco," but Kathleen leads me back into the house.

I naturally assume that we will work with neat little salt and pepper shakers and bottles of powdered nutmeg lined up on the table ready for our use. Wrong again. In the low light of the kitchen, I can make out the table covered in odd bowls, sticks, cones and scraps of unidentifiable objects, as well as the mortar and pestle.

In an aha moment, I realize that I am expected to prepare spices the seventeenth-century way: by grinding them with that mortar and pestle.

I pick up the ancient artifacts and examine them more as antiques than utensils. The mortar is the bowl, this one made of stone. The pestle is a wooden tool with a rounded end that's used for crushing and grinding spices in the mortar. The mortar-and-pestle combo has been used in cooking since 35,000 BC to today. It's just never been used by me.

Good, I think. This will be a new, fun experience. I may never have used a mortar and pestle, but I've always liked the way they look in old kitchens. I am adopting a positive attitude. I can hardly wait to show off the hidden power in my little biceps.

I love ginger, so that's the first ingredient I pick to grind down. The kind the Pilgrims used was dried and looked like small scraps of paper. I place a piece in the rounded bottom of the stone mortar and begin to grind it with the pestle. Grind, grind, grind. After a minute, I stop and look into the bottom of the bowl, expecting to see three or four tablespoons of perfectly ground powdered ginger.

But nothing has happened. The sheet of dried ginger is exactly the same as when I placed it into the pestle. I throw up my hands and exclaim, "What am I doing wrong?" Partly in disbelief and partly in frustration, I press down even harder with the pestle. After about ten minutes of excruciatingly hard grinding, I have produced a lot of perspiration and about half a teaspoon of ground ginger. Well, ginger fibres. It isn't even powder. Truly, all I can think is, "Where can I find a Cuisinart?"

I continue with peppercorns and then chunks of salt. The process dictates that I grind one spice at a time. Each one must be ground down, requiring almost as much muscular effort and mental determination as the ginger. Using the mortar and pestle is an exercise in patience and endurance.

The mortar and pestle was not the only device the Pilgrims used for attacking spices. I grind beads of nutmeg into powder with a small file. The sharp edges and small holes of the file are well suited for grating hard spices like nutmeg and mace. The only rewards for all my toils are the scents coming from the peppercorns (like a pungent juniper) and nutmeg.

I scrape a piece of dried bark from the Madagascar cinnamon tree with a knife to extract cinnamon powder, but again a lot of effort goes into producing not much powder. With my arms feeling like strands of limp spaghetti, I wipe the sweat off my face and step back from the table to admire my array of spices. We are, after half an hour, ready to add them to our dishes.

Kathleen moves to the fire and bends down to remove the lid from the pot. Inside, the duck is boiling in what is now a fatty broth. She drops into the broth the turnip I have chopped, adds salt and pepper and a sprinkle of my nutmeg, then replaces the lid for further cooking. I do a quick baste on our quails, which are turning golden brown and sizzling on the spit. I am ready to sink my teeth into a quail right now, but I hold myself back. Kathleen adds more wood to the fire to keep the heat consistent.

The next step is to fry the chunks of pumpkin over the fire. Kathleen reaches for a frying pan with a handle that is about three feet long. She places a dollop of butter in the pan. At first the elongated handle strikes me as quite curious, but its purpose becomes clear when Kathleen places the pan right into the hot embers of the fire to allow the butter to melt. I can see that the long handle allows the cook to manoeuvre the pan without being scorched. Kathleen takes the frying pan off the fire and I plunk into it the chunks of pumpkin that I have lovingly peeled and chopped.

Kathleen sprinkles the pumpkin chunks with my hand-ground salt, pepper, cinnamon, ginger and nutmeg. Even though there are a lot of spices, I am skeptical about how the pumpkin will taste. My mind is so tainted by the smoke, darkness and hard work that I can't believe anything desirable can be cooked in such primitive conditions. Kathleen puts the pan back into the fire and the pumpkin begins to sizzle. Even with the pan at the far end of a long handle, my face feels scorched from the fire and I am covered in sweat. Let alone everything else I have done, the entire process of cleaning, peeling, chopping and frying pumpkin is draining. However, when I look over at Kathleen, who is bouncing and joking around the kitchen, oblivious to the fire and smoke, I stop my mental whining.

Pumpkins were crucial to the Pilgrims' survival through the winter months. Years before the landing of the Europeans, they were grown by Native Americans, and seeds have been found at archaeological sites throughout North and South America dating back six thousand years. Native peoples ate the entire pumpkin, including the flowers and leaves. The Wampanoag were garden-to-table environmentalists before their time.

Just before all the dishes are plated, Kathleen slices bread from a loaf, arranges the slices on a trivet and puts them close to the fire for toasting. She then crumbles bread pieces into the duck pot, telling me that bread crumbs were used to thicken sauces. The Pilgrims also used bread crumbs to make gingerbread cakes and cookies.

The only grain available to the Pilgrims was corn. The Wampanoag ground it between stones and made a bread that they laid directly in the fire as a method of baking.

Pilgrims made their own version by boiling then pounding the corn and pouring it onto the floor of the outdoor communal oven to make a chewy flatbread. First they would build a fire in the oven to heat the chamber. When the oven was hot enough—experienced bakers could judge the temperature just by reaching inside the oven—they would remove the logs, lay the bread in the oven and plug the opening. The bread baked from the heat given off by the hot walls.

By the mid-1620s the Pilgrims were growing a variety of grain crops and using wheat, rye and corn flour to make a heavy bread called manchet.

While the pumpkin chunks fry over the fire, I reflect on the Pilgrims' plight. It's true that life was hard in Plimoth, but they believed they had a future there. Life had not been good for them in England, and in spite of all their challenges, they felt they could do better in the New World. They stayed.

This is the story for so many immigrants. They want a better life for themselves and for their children. They are willing to put up with threats and poverty and even risk their lives for the hope of a better future.

After three difficult years, life did improve for the Pilgrims. They got better at fishing, trapping, hunting and growing crops. More colonists arrived, and by 1627 Plimoth had a population of about 160.

After three hours of washing, chopping, peeling, grinding, roasting, basting, boiling, frying and toasting, our meal is ready to be plated. Unlike the Pilgrims, I have been spared such preparations as hunting the animals, plucking feathers, gutting and cleaning carcasses and building a fire, but nevertheless my morning activities have left me with an incredible appetite.

Considering the ingredients and the simplicity of the recipes, I am not expecting much in the way of taste. In fact, I am expecting a series of very bland dishes consistent with the plain interior of the Pilgrim house. If Pilgrim life was devoted to work and a lack of enjoyment, why would their food be any different?

As she removes the lid from the pot of boiled duck, Kathleen vanishes behind a cloud of steam. The vapours waft over to my face.

Hmm . . . I suddenly perk up. The aroma of the cooked duck is surprisingly enticing. Kathleen slices off pieces from the breast and thigh and places the steaming meat on my plate. She surrounds the duck with chunks of turnip and ladles on the broth that glistens with pearls of duck fat. As a garnish, she places a bud of mace, a type of seventeenth-century nutmeg, on the duck.

The first morsel of meat proves all my preconceptions completely wrong. I am in shock even. The duck, braised to a melting tenderness, is delicious. The broth has thickened perfectly and the salt, pepper and nutmeg have transformed the dish into a delectable stew. Even the turnips, al dente and infused with duck fat, are exquisite.

Now that my taste buds have been awakened, I am anxious to try the roasted quail that has been dripping buttery fat from the spit and turned a deep golden brown. Kathleen strips the spit of the birds and places one on my plate. She pours the drippings into a gravy bowl and adds dried bread crumbs to thicken the sauce. She covers the bird with the buttery gravy.

I cut off a leg and take a bite. It's sensational. Salty, peppery, tender and smoky, the meat slides off the bone. I have never enjoyed quail—or any fowl—so much. The flavour is intense, and all that butter basting has made the skin golden brown and crisp. I use the crusts of the dense bread to sop up the remaining quail sauce. I'm ready to eat the bones. In partial disbelief, I reach over and spear another quail onto my plate.

But the biggest surprise is the pumpkin. I've tasted pumpkin pie but I have no other experience of the orange squash. These beautifully caramelized cubes of orange goodness turn out to be as flavourful as they look. I think, "Pumpkin, where have you been all my life?" Fried crunchy crust on the outside and soft, pillowy flesh on the inside, they introduce me to a new flavour sensation. The cinnamon, nutmeg, ginger, salt and pepper release a pumpkin party in my mouth. The intense chopping and peeling were certainly worth the effort.

As we munch, sip and otherwise devour our dinner, Kathleen shares with me her intimate knowledge of Thanksgiving.

By the fall of 1621, the Pilgrims had built seven houses, a meeting hall and three storehouses for food and other supplies. After barely surviving that first winter, they had much to be thankful for—not least the Natives, without whose generosity in sharing knowledge and resources they would have starved to death.

One insight I gained from Kathleen was that the Pilgrims and the Natives did not necessarily "love" each other as is sometimes portrayed in the popular imagination. In fact, they didn't even "like" each other. However, they needed each other to survive and they co-operated as allies. The Natives were able to fight off neighbouring enemies with the help of the Pilgrims, while the Pilgrims needed the Natives for their knowledge of hunting, fishing and agriculture.

Nevertheless, Governor William Bradford wanted to express his gratitude to the Natives. He organized a celebratory feast, and they were joined by Massasoit, chief of the Wampanoag, and ninety of his men for three days of feasting and entertainment. The story of the harvest celebration is about two very different communities that came together, invested in each other and became stronger through a combined effort.

And eating was at the centre of it. To this day, eating together is the starting point for reaching a common understanding. Offering food is a way to express friendship, gratitude, generosity and empathy. When words fail, we can express ourselves through food.

As our dinner comes to a close, I thank Kathleen. I am so grateful to her for sharing her knowledge of the Pilgrim experience, the Pilgrim kitchen and cooking techniques. She

has been a fabulous instructor, and I am so impressed with her historical culinary expertise.

The last thing I say to Kathleen is that from what I tasted at Plimoth, the Pilgrims ate food that is just as delicious—if not better—than the food we eat today. She smiles and gives me a big hug.

Recently, Franny and I had a couple over for dinner after they had suffered a death in their family. They were shocked and saddened by their loss and feeling low. However, they said that it had been a pleasure and a source of solace to eat the food that friends and family had brought to their house for shiva, the week-long mourning period observed in Jewish homes. They smiled as they described the sumptuous stews, lasagnas and meats they had enjoyed for shiva dinners.

In the darkest, most terrible times, food is a comfort. When people arrived with food, the couple felt loved.

Even in grief, eating makes us know that we're alive. This is what the Pilgrims must have felt during the first horrific winters at Plimoth. The Pilgrims were conservative people who were dedicated to God. The severity of their religion limited their pleasures in life. But they enjoyed eating. It might have been their greatest pleasure.

Pilgrim life was hard. But at least their food was tasty. I'm glad that it was a comfort for them.

Plimoth Plantation

137 Warren Avenue, Plymouth, Massachusetts

www.plimoth.org | info@plimoth.org | (508) 746-1622

Dear Franny,

Here are some quick notes from Plimoth Plantation for our perfect kitchen. You might think that I've lost my mind, but there are some things from 1627 worth thinking about.

Everything Pilgrim is wood. It adds texture and warmth to a kitchen of stainless steel and sleek lines. Recycled wood siding and tree-trunk tables add history and character.

Access outside to extend the kitchen. It's good to chop veggies in fresh air.

A vegetable garden. You haven't eaten turnip until you've cooked them just minutes after they've been pulled out of the ground.

Let's grow pumpkins! They're big, orange and taste good too. Who would have thought that the 1627 Pilgrims would inspire my creative cooking.

We can mix new with old. Contrast modern fittings with age, texture and lines from an earlier time. An antique frying pan on the wall or a big old cutting board on the central island. Wait. I'm getting old. Maybe I'm *antique enough.*

The Pilgrim kitchen took me away from everything I take for granted and gave me an opportunity to touch, smell and taste things that are real. Being in a kitchen where people had to put so much effort into cooking gave me a greater appreciation for the finished dishes.

Oh yes, the mortar and pestle thing? That was fascinating, but let's please keep the food processor.

XOXO

J.

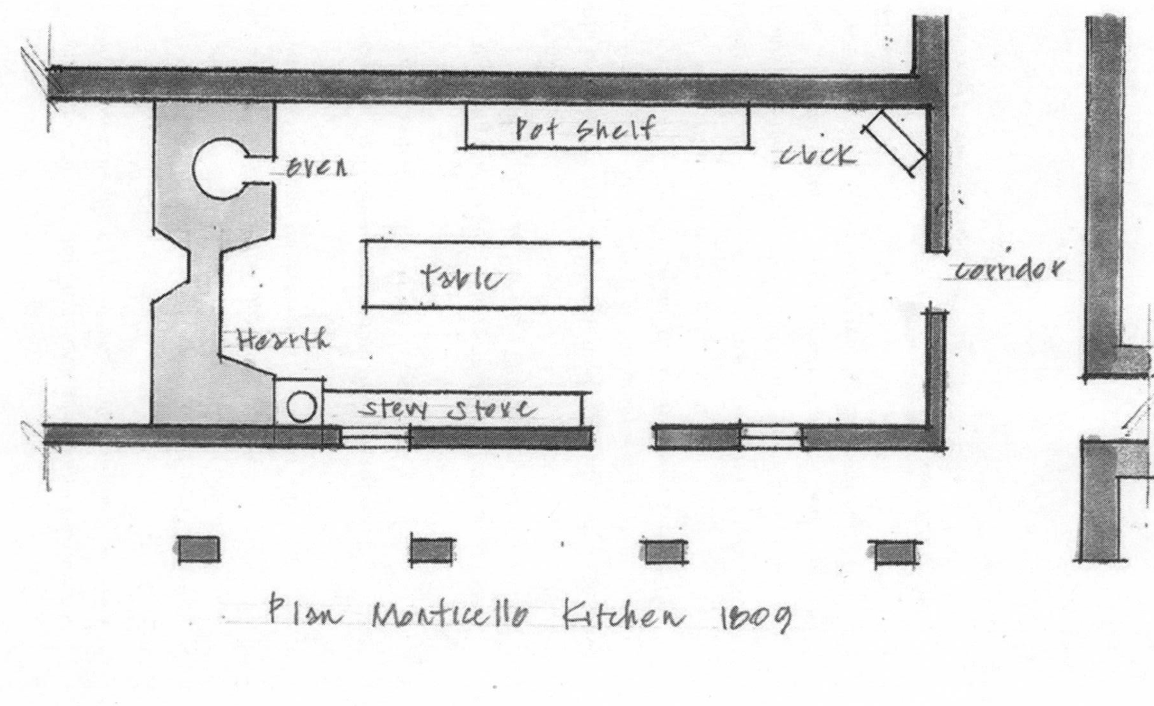
oven
Pot Shelf
clock
table
Hearth
stew stove
corridor
Plan Monticello Kitchen 1809

Interior Monticello Kitchen 1809

2

THOMAS JEFFERSON KITCHEN
at Monticello, 1809
Charlottesville, Virginia

If I could have dinner with any person in history, I would choose Thomas Jefferson.

I would ask him about his inspiration for writing the Declaration of Independence. I would ask him about the challenges of being the third president of the United States. I would ask him about his vision for his country and how that compares with what the country is today.

Most of all, I would ask him about his house, Monticello, about the renovations he was continually making, and the design of his kitchen. Jefferson is considered one of America's greatest architects, and he called Monticello his "essay in architecture."

When I studied the preservation of historic architecture at

graduate school, I became obsessed with Jefferson. I wrote papers on his buildings, I studied his designs and I read every book on him in the Columbia University library into the wee hours of the night. And so the kitchen I most want to see on my quest is understandably Jefferson's.

I was drawn to Jefferson because of his accomplishments as a nation-builder and designer of buildings. I was also drawn to him because he was supposed to be socially awkward. It is written that he did not fit the mould of other Washington politicians of his time. He was a poor public speaker and was slovenly in dress. As an insecure and impoverished young man at grad school, I was inspired by Jefferson's successes.

Jefferson was also a good writer. When the time came to write the Declaration of Independence, the Founders turned to him to compose a statement of the revolution and the reasons for their actions. And Jefferson delivered big time: "We hold these truths to be self-evident, that all men are created equal, that they are endowed by their Creator with certain unalienable Rights, that among these are Life, Liberty and the pursuit of Happiness."

I thought it outrageous that one man could write the Declaration of Independence, be a visionary for the country, serve as its president, be an inventor, botanist, scientist and astronomer and also arguably be America's greatest architect—one whose work reflects his vision for the country. When did the man sleep?

But as much as I admire Jefferson, I am also deeply disappointed by a contradiction in his life. Jefferson claimed to be philosophically against slavery, but it suited him to use enslaved people to operate Monticello. And though slavery in the late eighteenth and early nineteenth centuries was status quo in the southern states, I had always assumed that Jefferson was a decent man who treated his enslaved workers relatively well. However, recent portraits challenge that view.

If I could have dinner with Jefferson, I'd definitely bring up the subject—perhaps near the end of the evening. If I was going to be thrown out, I'd prefer it was after I'd finished eating.

When I heard about a cooking class at Monticello, I almost jumped out of my seat. It included a tour of the house *and* the kitchen, as well as the opportunity to cook and taste dishes that Jefferson would have enjoyed in his dining room. This was the closest I could ever get to having dinner with the man.

I told Franny that I had to go to Monticello. She said OK.

I had my bags packed in five minutes.

I arrive at the site of Monticello, close to Charlottesville, as the sun rises above the Virginia mountains.

Jefferson built Monticello at the top of a lofty hill to enjoy the views across the countryside, so I begin with a climb up a steeply rising road that cuts through a forest of poplars. I am soon short of breath. It must have been a challenge to transport food, supplies and building materials up this hillside.

After a vigorous fifteen-minute walk, I reach the crest and pause under the trees to enjoy the views of the hills, valleys and woods of Virginia. As the sun burns off the clouds hovering at the mountaintops, I understand why, during his time in Washington, Jefferson constantly yearned to return to this tranquility.

Admiring the scenery, I try to imagine what life was like here in 1809.

Monticello was the name of not only a house but of a whole plantation of crops in rolling fields, gardens, walkways and outbuildings spread over five thousand acres. Jefferson would have smelled smoke in the air from the kitchen and fireplaces. He would have seen an army of enslaved people working in the plantation fields, harvesting tobacco and, later, wheat. They also tended to his gardens with picks and shovels, digging up his crops, including enormous heads of red cabbage, twenty-seven varieties of kidney beans and his beloved Early Frame garden peas. There would have been hammering and sawing as carpenters worked on renovations. With 200 inhabitants, including 150 enslaved people; Monticello was a small village. There would have been a lot going on.

I see a small opening in the thick brush ahead, so I walk toward it. I follow the path and, through an arched opening in the linden trees, I spy a clearing and then beyond it . . . there it is. Monticello. A noble brick house that stands proud, bright and gleaming in the sunlight. I am thrilled to arrive at the house of Thomas Jefferson.

I stand on the lawn in front of the east entrance and slowly take in the broad facade. The central feature is the white portico porch and pediment that is supported by four Doric columns. I notice that people inside are raising the blinds on the three arched glass doors under the porch that lead into the house. On each side of the entrance are symmetrical wings accented by a series of tall windows with green shutters. There's a fanlight in the pediment, a weather vane on the roof and a clock over the front door.

From my studies, I know that Jefferson hated English Baroque buildings, which he felt were overly ornamented and reflected an aristocratic society. From books on the Renaissance architect Palladio, Jefferson absorbed a sleek, unadorned architecture of columns and pediments, a style that he thought promoted

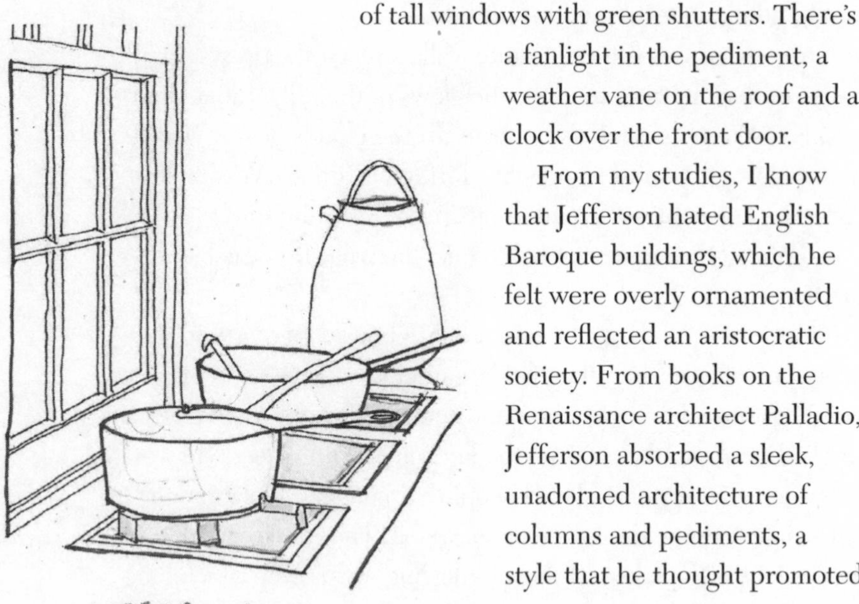

his egalitarian vision of the new United States. In contrast to the great manor houses of Europe, Jefferson designed Monticello as a modest, stately house appropriate for a gentleman farmer—not a king. For Jefferson, Monticello is his ideal of the American house.

Even though it is early morning, I am already thinking about cooking Jefferson-era recipes. I have been in contact with Eleanor Gould, who is the curator of gardens at Monticello and who will be teaching me to cook the kinds of dishes that were served at Thomas Jefferson's table. She has notified me that the kitchen inside Jefferson's house is a museum exhibit, so she will be cooking in a modern kitchen in the Monticello education centre. I will meet up with her later in the afternoon to indulge in some of Jefferson's favourite dishes, but first I want to tour this famous house.

I hold my breath and step through the arched entrance doors of the east porch of Monticello. As I enter, I immediately know that I am in the presence of a visionary mind.

I can see that this is not just an entrance hall but also a museum of Jefferson's diverse interests. I walk around the double-storey space and admire the Native American artifacts, European art, fossil bones, moose and deer antlers, animal pelts, sculptures and maps of Virginia and Africa. There is a welcoming spirit to the house.

A light bulb switches on in my head and I remind myself of my to-do list at Monticello, which is as follows: (1) watch where I am walking (2) don't break anything.

"Right," I say to myself.

I stroll straight ahead into the parlour, the social centre of the house that is appointed with furniture from France, multiple portraits and tall windows with views across the lawn. And I am in awe as I wander into Jefferson's bedroom and library, an intimate sunlit apartment with arched doorways and windows—

his nineteenth-century man cave where he sought solitude to write, draw and read.

But what impresses me most are the number of gadgets in the house. Among them are—

- The Great Clock over the front door that marks the days and hours with a pulley system of cannonball-like weights—a Jefferson invention.
- Automatically opening double doors. Jefferson hid a wheel-and-chain mechanism in the floor so that, at the touch of a door, both doors swing open like the supermarket doors of today—a Jefferson invention.
- A polygraph, a device consisting of a drafting board with two pens used to make simultaneous duplicates of the letters he wrote. It looks like a wooden lobster.
- A revolving bookstand that opens up to hold five books at adjustable angles. It allowed him to read five works at the same time—a Jefferson invention.
- A portable lap desk on which he wrote the Declaration of Independence in 1776. It looks like a twenty-first-century laptop—a Jefferson invention.

I add to my mental list of Jefferson's achievements that he seems to have been America's ultimate DIYer.

I step into the dining room, where I am surrounded by walls painted a brilliant chrome yellow, with tall windows and sliding glass doors to the tea room. The fireplace mantel is graced with books, a mantel clock and another Jefferson invention—two dumb waiters concealed on either side of the fireplace to carry bottles from the wine room in the cellar.

This Monticello house tour has been fascinating, but now I need to find the kitchen.

—

I locate a staircase that is barely two feet wide and head downstairs. It feels like I'm in a dollhouse. I imagine Jefferson's enslaved people struggling to carry buckets, brooms, hot water and platters of food up and down the narrow confines.

At the bottom of the stairs, I stand in a poorly lit and unfinished basement. I begin to walk a long, dark passageway, four feet wide, with whitewashed stone walls and a slate floor. At the end of the tunnel, I step into a burst of sunlight and gusts of fresh air. Before me is a panoramic view of Jefferson's vegetable garden, which stretches across the rural landscape. Hundreds of plants are divided into immaculate rows and terraced into the red clay hillside.

Outside the passageway, I am amazed to see built into the hillside a long, one-storey brick structure. I poke my head into the first room and see a large fireplace with a spit, kitchen utensils hanging from the wall and shelves of gleaming copper pots and pans.

In a building separate from the rest of the house, I have finally found Thomas Jefferson's kitchen. I have stumbled upon what Jefferson called the "dependencies," or service rooms. As with many houses of the time, Jefferson located the kitchen away from the main house in case of fire. Inspired by the drawings of Palladio, he built two underground wings into the slope of the hill to conceal the kitchen, pantry, beer cellar and wine room and connected them with a passageway in the cellar.

I step into the twelve-by-eighteen-foot kitchen. I admire the sunlight that streams in from the two windows and two doors onto the whitewashed brick walls and brick floor. Jefferson gave the space a clean, no-nonsense look. Designed to be simple and utilitarian compared with the rest of the house, it is roomy enough for several people to work in to prepare the meals, from a simple breakfast to lavish dinner parties.

A sturdy wood table in the middle of the room is laden with melons, pumpkins, freshly dug-up carrots and multicoloured gourds. Franny, who's a gardener, would be especially attracted to the basket of sculptural heirloom tomatoes sitting among the array. With their ridges, deep lobes, flattened shapes and bright red hue, they appear so ripe as to be bursting out of their skin.

The pride and joy of the kitchen is a long brick box built to waist height along one wall. It's Jefferson's stew stove.

Jefferson was apparently a great trader of seeds. I'm now wishing I'd brought some from Canada to swap with seeds from these tomatoes. The Monticello staff might well be interested in growing a few sugar maple trees.

At one end of the kitchen is a large brick fireplace. The hearth is equipped with a mechanical jack attached to a spit for rotating meat in front of the fire. Jefferson's cooks could roast whole sides of beef on it—helpful when feeding the many at his parties.

His cooks used the open hearth in other ways, such as boiling potatoes in pots on swinging cranes and frying fish on skillets set over trivets in the fireplace. I walk closer to the hearth and swing the metal hardware back and forth. The cooking temperature could be adjusted by moving the crane in or out of the fire or by adding another log. In addition, the fireplace has a built-in oven for baking bread with radiant heat.

But the pride and joy of the kitchen is a long brick box built to waist height along one wall. It's Jefferson's stew stove, a French kitchen appliance that was rare in the United States at the time. I can only think how much the Pilgrim women at Plimoth might have loved a waist-high stove to lessen the amount of up-and-down bending over their open cooking fire on the ground.

From this wonderful stove came many of the fine French dishes that became legend at Monticello. The cook would place charcoal in one or more of the eight holes in the top, and then cover the holes with grates that allowed the cook to control the heat under the pans. This fine control was necessary to make the delicate sauces called for in the French cuisine Jefferson favoured, such as sauce hachée, made with onions, shallots, meat stock, butter and mushrooms and served with red meat, or chicken fricassee, made with butter, flour, milk, cream, white wine and seasonings. Jefferson's favourite was made from onions, butter and cream to complement roast duck.

I can picture Jefferson's head cook, Edith Hern Fossett, and another cook, Fanny Gillette, standing over the stew stove and stirring the pots with wooden spoons. According to an article by Justin A. Sarafin in the book *Dining at Monticello*, the women travelled with Jefferson to Washington to be taught by a French chef in the President's Mansion (what we now call the White House).

When Jefferson travelled to Paris in 1784 to spend five years as United States minister to France, he did not expect to like the big city. He was a Founding Father who espoused an agrarian vision for America based on the attributes of the common, hard-working farmer. But as it turned out, Jefferson loved Paris. France not only stimulated his interest in architecture (plaques on classical buildings throughout southern France mark where he travelled), but it also opened up a new world of food and wine for him. On his return to America, he continued to enjoy French cuisine. He imported fine delicacies to Monticello from Europe, including Bologna sausage, Maille mustard, olive oil, figs, raisins, almonds, vinegars, oils and anchovies. He is credited with popularizing Parmesan cheese, french fries and Champagne in America.

Against the wall opposite the stew stove stands a free-standing shelving unit with a handy built-in countertop for preparing food. I'm transfixed by the gleaming copper pots and pans that stretch out across the shelving. Jefferson had these pots shipped back from France in 1790. (They were among eighty-six crates of goods.) Among them are a tart pan, fish poacher, chafing dish and kettle. Compared with cast iron, copper is lighter and has the superior heat conductivity essential to making those delicate sauces.

This kitchen would have been a flurry of activity from early morning until late at night with people carrying in wood to fuel the hearth; cabbages and potatoes hauled from the thousand-foot vegetable garden; apples and peaches collected from the orchard. There would be a continuous clatter of heavy iron tongs, shovels and pokers, and the banging of iron and copper pots compounded by the sounds of the cooks cutting, chopping and talking. The aromas of baking bread in the oven and sides of beef roasting in the fireplace would fill the air.

Jefferson's daughter Martha Randolf determined the dinner menu and was in charge of the kitchen. Like most Americans of the time, the family ate two meals a day, breakfast and dinner. Breakfast was served at eight o'clock and might include Monticello cornmeal muffins, hot wheat cereal, cornbread, butter and cold ham, with tea or coffee. Dinner (more on that later) was served mid-afternoon, and a light snack was made available for family and guests in the evening, with bread, cold meats and fruit left out on the dining room table.

I stretch my neck out the kitchen door and peer down the passageway to trace the path that the finished dishes would have taken from kitchen to the dining room in the house. It is a fair distance away. Jefferson disliked having servants present at meals, so he designed the kitchen and the path to the dining room to minimize contact with the staff. Enslaved people carried dishes in warming trays filled with hot water through the passageway underneath the house.

A staging area for plating was located at the bottom of the stairway that leads up to the main level. This staging area was also used for storing linens and crockery. Here the servants arranged the food on serving platters and added garnishes such as basil leaves or rosemary from the garden. Once the dishes were given their finishing touches, the servants carried them up the small staircase into the narrow passage that connects to the dining room.

I know I would be a complete clown on that tight stairwell, spilling dishes left and right.

Jefferson spent almost no time in the kitchen, but he took serious interest in the way food was grown and prepared at Monticello. The dishes from his kitchen more than simply entertained his guests. Jefferson believed that bringing opponents together over food and wine allowed them to better understand each other. He used food as a way of finding common ground.

It is mid-afternoon, and I am indeed very hungry. I am more than ready to cook recipes from Thomas Jefferson's table.

I make my way down the hill to the kitchen in the education centre, where I am warmly greeted at the door by my host, Eleanor Gould.

"Welcome to Thomas Jefferson's Table!" she says. Her big smile and enthusiastic manner assure me that good eating is soon to come. Also, the room is filled with the wonderful aromas of home cooking. This always makes me cheerful.

I seat myself at Eleanor's table and scan the heaps of vegetables, which include cucumbers, gourds, fingerling potatoes, lettuce, three varieties of heirloom tomatoes and bottles of vinegars. I can hardly wait.

We are in a state-of-the-art cookery classroom, with a demonstration table and stove at the front of the room and tables and chairs set out school-style. Floor-to-ceiling windows in the back allow for views of the lush Monticello grounds.

Eleanor begins by making a salad dressing from an 1824 cookbook called *The Virginia House-wife; Or, Methodical Cook*, by Mrs. Mary Randolph. "You really can't understand Jefferson's kitchen without understanding his garden," says Eleanor. "Even though mutton, pork, beef and fish were available, Jefferson loved vegetables. He wasn't a vegetarian, but he was unusually moderate in his consumption of meat. To Jefferson, meat was little more than a condiment."

Monticello was America's first experimental farm, and the thousand-foot-long-garden became Jefferson's laboratory. He kept detailed records of the growth of 330 varieties of fruit and vegetables in order to determine the hardiest and tastiest.

He made lavender flowers into ice cream and raspberries into vinegar. He'd be so on trend today.

Jefferson collected plants from around the world, among them sweet potatoes, Cherokee corn, peanuts, lima beans and sea kale. He cultivated white and purple eggplants—the white eggplants looking like duck eggs, hence the name eggplant. He knowingly broke the law when he smuggled unhusked Piedmont rice out of Italy in his pockets. He made lavender flowers into ice cream and raspberries into vinegar for salad dressing. He'd be so on trend today.

Eleanor passes me two small freshly picked heads of Tennis Ball lettuce and prompts me to wash them. They have broad curly leaves and resemble present-day Boston lettuce.

Eleanor asks me to start the dressing by mashing up two cooked egg yolks with a wooden spoon in a bowl and add some olive oil.

I then add salt, sugar, Dijon mustard and a vinegar made with tarragon from the garden. I give the dressing a gentle stir with the wooden spoon. As Eleanor pours the dressing over the bowl of lettuce leaves, she explains that we have made a historic vinaigrette dressing that would have been a mainstay on the Jefferson table.

"Jefferson loved salad and planted lettuce every Monday," she tells me. "He was a man with adventurous tastes and had people sending him seeds from around the world."

My next job is to toss the Jefferson salad with wooden tongs, and then it is time to taste it. Fork in hand, I pause. I am skeptical of this old-time dressing made with sugar and egg yolk. But as I crunch down and my tongue absorbs the mustardy tarragon tartness, all my doubts vaporize into the Virginia afternoon air. This salad would be most welcome on any French restaurant table.

After I devour the first bowl of salad, Eleanor offers me a second helping.

"Really? Thank you, Eleanor. Don't mind if I do."

While I graze on my second serving, I daydream of how a dinner at Monticello might unfold. I think back to an essay I read by curator Susan Stein in the book *Dining at Monticello* in which she describes a Jefferson dinner party.

At Monticello, the main meal of the day took place at three in the afternoon, when butler Burwell Colbert would ring the dinner bell and swing open the dining room door. Guests would be greeted by Thomas Jefferson and his daughter Martha Jefferson Randolph, who managed the household. If I found myself at dinner, I would tell them both how honoured I was to meet them and how much I loved the house.

The dining room table would be covered with a bountiful display of dishes for the main, or "first," course. Dinner was served in the French style, meaning guests served themselves rather than being served by the host or servants in the manner of English service. The day's menu might look like this:

Rabbit Soup
Okra Soup
Vermicelli Soup

Mutton Chops
Baked Virginia Ham
Beef à la Mode
Beef à la Daube Lemaire
Creamed Cod

French Beans
Cauliflower in Brown Onion Sauce
Multiple salads

Beer and cider

At the end of the first course, servants would clear away the plates and the tablecloth. They would then arrange the second course—trifles, cakes, tarts and such—symmetrically on the table. I love dessert, and if I were there I would keep an eye open for a spectacular-looking treat called Snow Eggs, attributed to James Hemings, one of Jefferson's French-trained cooks. It was composed of fluffy white clouds of soft meringue floating above a yellow custard and garnished with ruby-red raspberries.

After dinner, Jefferson would escort guests to the adjacent tea room for refreshments. Jefferson would enjoy a glass or two of wine after dinner. Wine was a passion of his, and he is considered by many to be America's first great wine aficionado. If I were fortunate enough to be present, I might help myself to the nuts and sweetmeats on offer in little baskets hanging from an elegant glass tree, or epergne. When it comes to food, I like to try everything.

I feel a tap on my shoulder. It is Eleanor, rousing me from my dinner-party daydream.

"John, I'm sorry to bother you, but there is still work to be done," she says.

She presents me with a basket of red heirloom tomatoes and cucumbers, and I get started slicing the vegetables and laying them out on a china platter. Eleanor tears up basil leaves to garnish the vegetables and then pours on olive oil that gives the salad a glistening final touch. Jefferson was a pioneer grower of tomatoes, and his family kept recipes for pickles, preserves, omelets and ketchup.

As I bite into a bright red tomato, my mouth is filled with an explosion of fruity sweetness. The nutty olive oil and tangy basil enhance the fresh flavour. Simple, but so classic, the Jefferson salad reminds me that feasting on just-picked tomatoes is one of the ultimate pleasures of summer life. I take my time with each fabulous forkful.

Eleanor brings out a large porcelain terrine. It contains a magnificent tribute to Jefferson's garden. A cloud of steam rises as she removes the lid to reveal a colourful hot gumbo made with carrots, lima beans, onions, tomatoes, squash, parsley and thyme. Its secret ingredient is okra, introduced to the Monticello kitchen by its African cooks. It thickens the soup and imbues it with an earthy flavour.

It is a light but tasty gumbo, about to be given extra zest courtesy of a few drops from a bottle of Texas bird-pepper vinegar. Jefferson obtained the seeds in 1812 and was soon incorporating the bird pepper into spicy sauces, vinegars and pickles. It has a definite kick and sends an immediate rush of heat to my face. I like hot and spicy, so I give the bottle a couple of extra shakes.

I am filling up on gumbo, salad and tomatoes when Eleanor removes a magnificent pan of bubbling macaroni and cheese from the oven. Pasta from Italy—Jefferson called any shape "maccaroni" (sic)—became a favourite dish of his while he was in Europe. He took notes on the type of flour used, how the dough was pressed, and even had a macaroni form shipped to him from Italy. Jefferson was not the first to introduce macaroni to America, but he is credited with making it popular. When he served macaroni and cheese at a state dinner in 1802, it became the talk of Washington.

Eleanor passes me a knife so that I can cut thick slabs from the pan. They are rich and decadent-looking, with a delicate crust on top. This is no ordinary mac and cheese. Eleanor tells me, "The recipe includes a custard of butter, cream, dry mustard, eggs and extra-sharp cheddar cheese that was poured over the macaroni noodles and set in the oven."

As I attack my slab and let each mouthful roll around and melt, I feel a sense of pure pleasure. The cheese, cream and butter custard melds perfectly with the chewiness of the noodles. "This is way over the top," I say to Eleanor. "This is easily the richest, most luscious mac and cheese I have ever eaten in my life."

Once again, my expectation that historical food would be bland and tasteless is completely mistaken. This is more of an intense quiche with noodles than a mac and cheese. My bowl of salad with the mustard vinaigrette is a perfect light accompaniment.

While we dig in, Eleanor tells me the story of the Cheshire Mammoth Cheese. This was a gift to Jefferson when he was president from the people of the town of Cheshire, Massachusetts.

The cheese was four feet wide, fifteen inches thick, weighed 1,234 pounds and was made with milk from every cow in Cheshire. Too huge to be transported by wagon, it was slid by sleigh to the President's Mansion in Washington during the winter. The three-week, five-hundred-mile journey caused a sensation. Citizens came out to celebrate the cheese at each stop along the way.

Eleanor explains that the cheese was as much a political statement as a present. It was made by farmers and their wives without any help from enslaved people, and the president was made aware of that fact. It became an attraction at the President's Mansion, but after two years Jefferson and his dinner guests had completely devoured it.

I am now so happily stuffed that I feel like *I* need horses and a sleigh to pull me away. But the cooking class is not finished.

"It's time for dessert," announces Eleanor. "Always a special course at Monticello. We are about to make one of Jefferson's favourites—profiteroles."

I can't believe my good luck. As a great lover of cream puffs, I suddenly feel slightly less full. How can I say no? I must indulge in the name of historical research.

Gliding over to the stove, Eleanor heats milk, butter and salt in a small pot over a medium heat. She adds flour and eggs to the liquid, and as she continues to stir, the mixture magically turns into a dough. I have never seen dough made over a stove. I am even more fascinated when Eleanor proceeds to spoon the mixture into a pastry bag and squeezes it out into mini swirls on a baking pan. After making a half-dozen of the little flowerets, she offers me the pastry bag.

"Would you like to give this a try?" she asks.

"I used to make curvy ice cream cones at a Dairy Queen as a teenager," I tell her. "This couldn't be much different."

Even though it has been many more decades than I'd like to admit since I worked at the DQ, I lean over the counter. With one hand I hold the top of the pastry bag closed and simultaneously squeeze it while with my other hand I guide the circular motion. While Eleanor watches over my shoulder, I feel a quiet sense of accomplishment as the dough flows out of the bag and onto the metal pan. It takes a little while, but once I get the hang of it, my flowerets look similar to Eleanor's, albeit a little larger, lopsided and artistically . . . abstract.

"Very good, John," says Eleanor. "You must have been a valued employee."

Before I know it, presto—we have finished off the pastry bag and made a tray of profiterole-dough flowerets. I pop our baking tray into the oven and we sit back to enjoy the aroma of Jefferson home baking waft through the room.

In Jefferson's time the preparation of puddings, pastries and trifles was under the supervision of the lady of the house, in this

case Jefferson's daughter Martha Randolph, who sometimes also cooked the main meal. She kept her eye on things because sugar was expensive and the stream of hungry visitors to Monticello added to the family's money woes. Jefferson went deeply into the red, thanks to his generous hospitality, fine living and the renovations he was constantly making to Monticello. In the end, the former president and Founding Father died broke.

After about twenty minutes of conversation and waiting, Eleanor pulls out a bowl of ice cream from the freezer and plunks it on the kitchen counter. She pulls out the baking sheet of profiteroles and, to my delight, they are an irresistible golden brown. Eleanor begins to slice off the top half of each with a sharp knife. As steam rises from their insides, Eleanor scoops in dollops of ice cream. The ice cream immediately begins to soften and melt, covering the pastry with a thick flowing custard.

In Jefferson's time, dinner guests marvelled at being served a dessert that was hot and cold at the same time.

Jefferson is sometimes credited with introducing ice cream to America, but that is a myth. But he did promote ice cream and served it frequently at the President's Mansion. Later, he had an ice house built at Monticello, where he stored ice cream along with butter and fresh meat. Further evidence of Jefferson's love of this frozen dessert are his handwritten recipes and Monticello's collection of moulds and an ice cream maker.

Eleanor places the dainty profiteroles on a plate and garnishes them with slices of Lemon Cling peaches. Peaches were another of Jefferson's passions. He grew thirty-two varieties at Monticello, including this one.

We sit down to taste our dessert. I cut into the profiterole with my spoon out of view. As I savour the creamy, hot/cold confection, I think about cooks who would have made it.

Visible or not, Jefferson's cooks turned out sumptuous French dishes in his kitchen that were widely praised by diners at Monticello. Their thoughts on the matter have not been recorded. But culinary history does record, with deep respect, the names of Edith Hern Fossett, Fanny Gillette and James Hemings.

Monticello
931 Thomas Jefferson Parkway, Charlottesville, Virginia
www.monticello.org | (434) 984-9880

Dear Franny,

I think you always know a person better after you've visited their house and I feel like I have a better understanding of Jefferson—his inquisitiveness, creativity and his kitchen—after being at Monticello. And I think there are a number of pointers we can take from his house for our own kitchen.

Jefferson showed off his copper pots like valuable jewels. Pots and pans are part of the show—especially if they're copper.

A passionate gardener, Jefferson kept his vast garden right outside the door of the kitchen. We could too (though not as large).

Jefferson experimented with vegetables and seeds. We could have a greenhouse in the kitchen—just something to start seeds in the winter. It might help us get through our February blues.

Innovation was everywhere at Monticello. Maybe something like an eating counter that could swing/pivot/slide back and forth into the wall. Wait a minute . . . that wine bottle dumbwaiter could be fun too. Anything to make cooking easier.

We have to keep an eye on the money. Jefferson died bankrupt. Monticello is beautiful and pristine now, but during his time it was a constant worksite. We definitely need to stick to a budget.

There are contradictions in Jefferson's life. But there are contradictions in everyone's life. Even though I've made this visit to his house and walked the halls, experienced the kitchen and eaten the dishes, you would think that I've had enough of Jefferson. But no—I find my fascination has increased and I want to return to Monticello.

More than ever, I wish I could have dinner with him.

XOXO

J.

P.S.: How would you like some of that Jefferson copper cookware for Christmas? I'll look for it on sale.

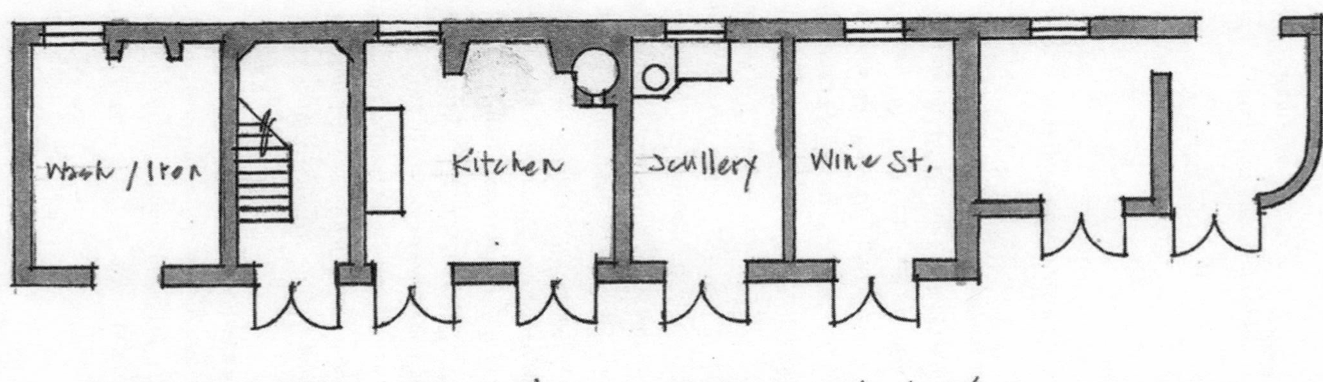

Plan Hermann-Grima Kitchen Building

Interior Hermann - Grima Kitchen

3

HERMANN-GRIMA HOUSE KITCHEN, 1831

New Orleans, Louisiana

The creamy seafood essence of shrimp étouffée. The pillowy softness of sugar-dusted beignets. And the burnt-sugar nuttiness of candied pecans. These are a few of my favourite things, and they all originate from the great city of New Orleans.

In fact, they originate from one kitchen here. It is located in the Hermann-Grima House, a legendary site of lively entertainment and French Creole fine dining that drew the upper classes of the city in the 1830s. Owned by Samuel Hermann, a successful cotton and sugar broker, and his wife, Marie Emeranthe Becnel Brou, the daughter of a wealthy planter family, this house was one of the most architecturally significant in the French Quarter, and it remains an outstanding historic building, even in a city spoilt for them.

But of course it is the kitchen that I have come to see.

The food, music, art and spirited people of New Orleans all lift my spirits, but it is the architecture that thrills me most. Thanks to the many organizations devoted to preserving historic buildings in New Orleans, the original quadrant, the French Quarter, has maintained its nineteenth-century scale and integrity. Its lacy wrought iron railings dance above the street, its brightly painted shopfronts beckon passersby off the sidewalk, and its Spanish Colonial architecture define the city's character.

Following fires in 1788 and 1794 that destroyed hundreds of wood-frame houses, the ruling Spanish government declared that any building of more than one storey must be constructed of brick. The resulting three-storey structures give a Mediterranean look to the Quarter. Designed with floor-to-ceiling shuttered windows to allow for maximum ventilation and sweeping balconies providing the interiors with shade from the sun, they have shopfronts at street level and residences above. Typically, an arched alleyway leads to a private porch or loggia in the rear, graced with a courtyard.

But today, I have extra incentive to visit the kitchen at the Hermann-Grima House. There is the promise of cooking and eating Creole delicacies over the open-hearth fireplace and potager stove with an 1830s cooking expert. I can almost smell the steaming shrimp and taste the tangy spices already. My hope is to inject some of the fun and pepper of this great city into our perfect kitchen.

I stride ahead on Bourbon Street and take a right onto narrow St. Louis. I am ready to cook.

The Hermann-Grima House is an aberration in the French Quarter streetscape. Despite its eight-thousand-square-foot expanse, one could pass it by without noticing. Designed with simple brickwork and restrained ornamentation, it is built in the plainer Federal style, and so it easily fades into the background. Compared with its more extroverted Spanish Colonial neighbours, preening and demanding attention, the Hermann-Grima House is the handsome southern gentleman that quietly blends in at the party.

Maybe it's the modest dimensions of the windows or the simple ornamentation of the wood shutters, but Federal houses like the Hermann-Grima mansion are some of my favourites in America. They are solidly built houses, without any unnecessary flourish. More typically exemplified by row houses in Boston and Brooklyn, the style was popular in America from about 1780 to the 1830s. It is named after the period when the Federalist Party dominated American politics. As Americans from the northeast migrated south, they brought their architectural tastes with them.

The brick-built Hermann-Grima House has three storeys, a centred front door and a symmetrical facade. Each of the eight windows in the facade comprises small panes. (Large pieces of glass were expensive at the time.) Most of the embellishments on the facade are concentrated on the front door: fluted Ionic columns, sidelights, a classic horizontal entablature and a Greek Revival fanlight. Derived from the classical Roman architecture of Palladio, fanlights were installed both to provide light in the hall and to act as an elegant showpiece of the house frontage. The semi-elliptical fanlight at the Hermann-Grima House spreads out like a sunburst.

I am drawn to a lush courtyard at the side of the house. I place my hand on the handle of an iron gate and push down experimentally. There is a click and the gate swings open. Feeling lucky, I look to the left, look to the right and step inside.

Now away from the madness of Bourbon Street, I sit on a wood bench and inhale the scent of trees heavy with oranges, lemons and kumquats. The leaves rustle, and an enormous lemon, the size of a softball, lands with a thud. The courtyard oasis is enclosed by the main house to the east, a more modest three-storey brick structure with outside walkways to the north and brick walls on the west and south sides of the garden. My eyes scan the distressed brick of the 1831 mansion, the weathered wood shutters and flagstone pavers dappled with sunlight. If there is a more beautiful and peaceful courtyard in America, I haven't seen it.

I sketch and take photographs until I am greeted by Jennifer Dyer, an education officer at the Hermann-Grima House and my cook for the day. Vivacious and congenial, she welcomes me with a firm handshake. Jennifer, a native of New Orleans, tells me that she has a master's degree in history from Southeastern Louisiana University and previously worked as an educator at the Beauregard-Keyes House and the Oak Alley Plantation.

"I love my job," she says. "I get to meet people and make dishes that have been passed down through the generations. We use the cooking methods and foods of early nineteenth-century Creole cookery. For a historian who likes to cook, it doesn't get better than this."

Her friendly and convivial manner gives me immediate confidence that this will be a great day of learning about two of my favourite subjects: New Orleans architecture and New Orleans food. (Jennifer Dyer, by the way, is no longer at the house, and since my visit there have been changes in the staff and management.)

Jennifer agrees with me that building a Federal-style house in the French Quarter was an unconventional architectural statement. "But Samuel Hermann wasn't a conventional person," she explains. "As a German-Jewish immigrant, he embraced the American Dream and was open to its possibilities."

Born in 1777, Samuel Hermann, like many others from Germany, settled along the Mississippi River between New Orleans and Baton Rouge on what became known as the German Coast. Developing a skill in business, he became a broker between cotton plantation owners and New Orleans merchants. In 1806, at the age of twenty-nine, he married Marie Emeranthe Becnel Brou, a Creole widow with two young sons. Emeranthe's family had been influential in the area for decades, and she was a part of the highest social circles of Louisiana society.

The 1800s were a promising time for entrepreneurs like Samuel Hermann in New Orleans. Because of the high demand for plantation crops, and the unpaid labour of hundreds of thousands of enslaved people, businessmen could make large profits. Thanks to the introduction of the first steamboat in 1812, Hermann could take advantage of the opening of the city's trade to the interior and plantations. Before the steamboat, freight was barged down the Mississippi River from Louisville, Kentucky, on a trip that took weeks. The new steamboats were larger and faster. New Orleans became the second-busiest port in the nation.

By the mid-1830s, the southern cotton market was booming and New Orleans was becoming the cotton capital of the world. Because of the long influence of the French in Louisiana—as its first settlers and colonizers—French was still the preferred language in the city, and the ten-dollar bill bore a *dix*. Spirited businessmen like Hermann followed their dreams in the land of "Dixies."

In 1823 Samuel engaged the services of William Brand, a Virginia architect, to design a Federal-style house. Although a departure from the overall style of the French Quarter, the Federal style of architecture was closer to the tastes of Hermann's eastern business associates as well as to the Greek Revival style of large plantation owners. Jennifer tells me that this was a house that would give a newly established and successful businessman like Samuel Hermann respect and differentiate him from the traditional wealth in the city.

When the Hermann house was being built, French Creole was still the dominant culture in New Orleans. Following the Louisiana Purchase of 1803 that transferred the region to the United States, an influx of Americans from the northeast arrived with ambitions in sugar planting and land speculation. Social and economic tensions grew between the established Creole people and the new Americans. They despised each other. The Creole saw the newcomers as being unrefined and brash while the northern Americans judged the Creole to be the snobbish old guard who were protecting their power. With a Creole wife and American clients, Mr. Hermann had loyalties to both sides.

By 1836 the rivalry had grown so bitter that the city officially split into three districts to keep the feuding residents apart. New Orleans was, and continues to be, a passionate city.

I am about to discover that, like Jefferson's Monticello, there is more to this house than meets the eye.

Jennifer leads me to a handsome three-storey brick structure, about twenty feet away across the courtyard. It is behind the mansion, completely detached and sitting beside a grove of citrus trees. Batten shutters span the ground level. On the upper two levels are eight shuttered doors facing onto exterior walkways.

"This is the kitchen building," says Jennifer. "I have to build a fire in the fireplace and do some chores, so why don't you check out the rest of the building?"

I step through the door closest to the mansion and find myself in a dark, damp-smelling room of peeling yellow paint, stucco and exposed brick walls. One wall is filled floor to ceiling with full wine racks. Large wooden barrels sit on the stone floor against the opposite wall. At the city market, Hermann bought casks of wine and liquor from France and Spain and had the contents filtered, bottled and stored in this room.

Next door is a scullery. Plain white plates are stashed on a wooden dish rack. There's a round sink, called a set kettle, with a wood-burning stove underneath that was used for both washing pots and cooking soups and stews.

I poke around the antique utensils on a side table and am intrigued by a wooden box for granulating sugar from solid loaves, a water cooler with a pumice filter, and a screw-down berry press resembling some kind of medieval torture instrument.

In a third room are washbasins, a central wood table and an iron and ironing board. As in any large household—especially one where table linen was changed after every course, in French service style—laundry was a constant chore. I step outside, full of sympathetic thoughts for the enslaved people who had to wash and iron all day in the city's oppressive heat and humidity.

Finally, I arrive at the fourth doorway, the all-important kitchen. As I step through the batten shutters, I am met by Jennifer.

"Welcome to Mrs. Hermann's kitchen!" she says. "You're just in time to help me make an 1830s New Orleans breakfast."

Intense heat from the fireplace hits me in the face like a Louisiana freight train. Inside the enormous hearth, leaping flames spit out sparks that crackle and pop in the air. It is a spectacular greeting. Jennifer detects an uneasiness in my body language. I guess it is my cowering from the flames and shielding my face with my hands that gives the game away.

She assures me that everything is under control. The kitchen was kept separate from the main house to prevent overheating the house in summer and to contain any outbreak of fire. In addition, a water well built under the courtyard could be used to help extinguish flames. A separate kitchen building also insulated the family and guests from the clatter of pots and pans and the smells of cooking.

A central table is covered with dishes of spices, bowls of vegetables and pitchers of milk that Jennifer has laid out. She tells me she was working till late last night, collecting and sorting nineteenth-century recipes for my visit. Her training in historical cooking methods is evident as she moves around the kitchen with the ease of a nineteenth-century master chef, manipulating strange-looking utensils and working at the open fire with uncanny calm and ease.

"People in New Orleans love to eat," says Jennifer. "Of course, you could say that about people in any city. But in New Orleans we have our roux, étouffées, the influences of so many cultures, a position as a port city and a culture of partying and entertaining. So it's especially true of us."

I couldn't agree more. When some people think of New Orleans, they think of Mardi Gras and Bourbon Street,

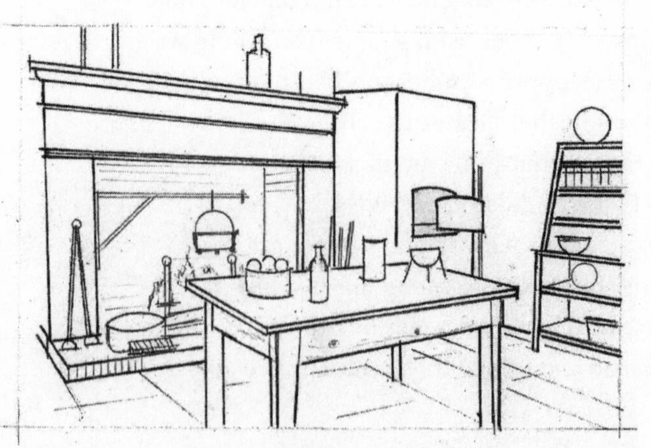

Hearth & oven

but to me, New Orleans is all about dark roux (a butter and flour mixture used as the base for soups and sauces), andouille sausage, cornbread and pralines. Like the famous gumbo, the city has incorporated all its historic influences—Spanish, French, American, Creole, Caribbean, Cajun, Italian and German—and melded them together to make what is a uniquely New Orleans cuisine.

I learn that the history of European cuisine in New Orleans begins with the French in 1682, when the explorer La Salle canoed down the Mississippi and named the region La Louisiane, after King Louis XIV. The French learned to use cornmeal to bake cornbread and to finesse French dishes with locally grown foods and herbs that the Native peoples introduced them to. But while the history of New Orleans is sometimes portrayed as a story of the French and Spanish repeatedly marching in and marching out, leaving behind recipes from their own cultures, Jennifer tells me that the development of Creole cuisine is more complex than that.

The magnificent sausages that I love in Creole dishes were produced by German families who were given free land to farm in Louisiana in 1717. Enslaved Africans brought to Louisiana in 1719 introduced their own recipes and ingredients—gumbo stew, okra—to their French owners. When the Spanish took control of Louisiana from France in 1764, they introduced the lively spices such as hot cayenne and bell peppers that are the signature of New Orleans dishes. Jambalaya, the peppery rice dish, evolved from the Spanish paella. And between 1764 and 1800, the colonists traded with Caribbean countries, which added Caribbean foods and influences to the ever-evolving New Orleans cuisine.

The happy result of this mix of flavours was Creole cuisine. In the early 1700s, the term *Creole* was used to distinguish the Louisiana-born French from newly arrived immigrants. Later, *Creole* was broadened to include any Louisiana-born people of French, Spanish or African descent or mixed ethnicity.

"In New Orleans, we say that the flavour of our past is still present in our meals," says Jennifer.

Standing in the Hermann-Grima House kitchen, I am surrounded by antique iron pots, pans, bowls and curious-looking utensils. I have to wipe the sweat off my face as the fire grows even more intense and fills the room with smoke. Jennifer points out the controls for the fire—a wet wool blanket and a water bucket.

Facing onto the tranquil courtyard, the twelve-by-fifteen-foot space is large for its time. The walls—yellow stucco parged over brick walls—are clean and unadorned. It has a fourteen-foot ceiling, and batten shutters serving as doors to the exterior to allow for ventilation. The flagstone pavers of the courtyard extend inside as a continuous floor, making a nice visual connection. At first glance, it appears to be a basic space for cooking. But as I look closer, the Hermann-Grima kitchen reveals itself as anything but basic for 1831.

The Hermanns entertained often, as evidenced in several surviving letters written by their contemporaries. Eating and drinking in the French style were central to their professional and social status as an upper-class New Orleans family. To reinforce her position in New Orleans society and to enable her husband to regularly entertain his business contacts, Mrs. Hermann would have wanted to host memorable parties. To do so, she would have needed a state-of-the-art kitchen.

She got it. This is surely a luxurious kitchen for its time. It could cook for gatherings of up to three hundred people.

- Luxury #1: Large hearth. In the hands of a skilled cook, the large hearth—the sole source of heat—was capable of accommodating a twelve-course dinner.
- Luxury #2: Plentiful wood for cooking. Firewood was scarce in New Orleans, and large supplies of lumber were mostly reserved for shipbuilding on the waterfront. Fireplaces

and stoves typically used coal shipped in from Kentucky and West Virginia. A steady supply of (expensive) firewood, however, was not beyond Mrs. Hermann's means.

- Luxury #3: Beehive oven for fresh bread. Even though bread was available at the New Orleans market every day, Mrs. Hermann wanted her bread baked in her own oven.
- Luxury #4: Potager. The real star of the show in the Hermann-Grima kitchen is the four-hole potager, a stovetop comprising four "stew holes" with variable heat control. Potagers were rare in early nineteenth-century North America, but for the Hermanns—as for Thomas Jefferson—heat control was essential in making the delicate French sauces featured in fine French cuisine. Smaller than the Jefferson eight-hole stew stove, the Hermann kitchen's potager is a long, narrow masonry box raised to waist height, like a kitchen counter, with several holes for saucepans. To build heat, there are holes on the front in which the cook lays a fire or loads coals from the main fireplace. Grills can be placed inside or above the hole, depending on how much heat is wanted.

Another "luxury" must be noted: the enslaved people who cooked and served the food. Fifteen worked for the Hermanns.

Jennifer gets ready to prepare a light breakfast of pain perdu, or French toast, and sausages.

In the 1830s, she tells me, breakfast was served to the Hermanns at ten in the morning. Menus were seasonal, so hot breakfasts were served in the winter and lighter, cool foods were served in the hot summers. As she collects the heavy cast-iron pots and pans, she explains that the breakfast will be cooked exclusively in the hearth, over the open fire. She arranges the utensils around the fireplace and lays out all the ingredients on the central table.

She begins by beating eggs in a bowl, then pours in milk from a pitcher to make a custard. The custard is the same as we use today to make French toast. With the open-hearth cooking method, after dipping the day-old bread slices in the mixture, they are placed on a round greased griddle. The griddle is on a swinging arm—called a crane—over the fire. As the toast is frying, Jennifer takes a shovelful of hot embers from the fire and spreads them at the foot of the fireplace. She puts sausages on a small iron grill over the embers, and soon I hear the wonderful sound of sizzling.

Preparation for the household's daily meals would have begun at daybreak, when two or three servants walked to the New Orleans market for fresh produce and meats. With no refrigeration and a hot climate, they would have bought only the groceries that were needed for that day (or would keep at room temperature). The New Orleans market opened in 1791 and still stands on the same site today.

As the second-largest port in the United States, New Orleans had access to vegetables, fruits, meats, wines and spices from all over the world, along with local produce, meats and fish. It is one of the reasons there is such a great food culture here.

Diversity ruled, and shoppers could choose from the assortment of French, Italian, Spanish, Portuguese, Dutch and German vendors. African Americans sold coffee, pralines and rice fritters, and the native Choctaw brought herbs, spices and crafts. Citizens gathered at the market to shop, gossip, drink coffee and add to the collage of colours, nationalities, languages and culinary traditions of New Orleans.

"One day I was making a pineapple dish in the kitchen," Jennifer says. "A visitor thought they wouldn't have had pineapple here in the nineteenth century, but in fact they did, thanks to New Orleans being a port city."

The French toast begins to brown at the edges, and Jennifer deftly flips over the slices on the griddle.

Mrs. Hermann's role was to ensure that the house ran smoothly, while the actual work, of course, was done by servants. Mrs. Hermann designated one to be in charge of the house and one to be responsible for the kitchen. Typically, three servants would work together in the kitchen to cook the elaborate multi-course meals. Servants carried the finished food in covered cooking pots across the courtyard to the main house.

Working in this kitchen required strength. I pick up the waffle iron. It's a two-sided clamp with a long metal handle. It feels like a barbell from the gym. It must weigh fifty pounds. I cannot imagine lugging this hunk of iron around for any length of time, never mind making waffles with it after it has been heated in an open fire. The job here was part cook, part foundry worker.

Jennifer plates up the meal on the work table. She generously dusts the French toast with powdered sugar and heaps sausages onto the plate. She finishes the dish with slices of satsuma, small, loose-skinned mandarins that have just been picked off the trees in the courtyard, and cucumber pickles that have been in a brine of garlic and vinegar like the ones the Hermanns' staff would have made.

I pick up the waffle iron. It's a two-sided clamp with a long metal handle. It feels like a barbell from the gym.

"There was a preference for strong flavours in New Orleans cuisine—sweet, sour, spicy," says Jennifer. "It would not be unusual to put several teaspoons of sugar in a cup of coffee."

Sugar cane was a boom crop in Louisiana. The soil supported sugar cane and the plantation owners profited from slave labour.

I press my fork down into a piece of French toast. The outside has a lovely crunch, while the inside is soft and fluffy. The egg custard and butter from the griddle meld beautifully with the powdered sugar and satsuma. A sweet-and-sour contrast is created by the garlic pickles. I marvel that such delicious food can be prepared over an open fire with what looks like medieval cooking implements. It is perfect French toast. I can only think that whenever I have it in the future, I'll be saying to the plate: "French toast, we always had New Orleans."

The last morsels of heavenly pain perdu disappear into my mouth, and then I help Jennifer clear dishes. She tells me that she has to get things together to make the main meal of the day.

"I think this is a good time for you to tour the main house," she says. "Enter by the front door to get a feel of what it was like to be a guest of Mrs. Hermann. When you come back to the kitchen we can start cooking again."

As I stand at the front door outside the Hermann-Grima House, the first thing I notice is how all the neighbouring three-storey houses are built tight up to the edge of St. Louis Street, with no setback. With only the separation of three steps down from the front door and a narrow sidewalk, the door basically opens right onto the street. No garden, no strip of sod. This immediacy of the entrance combines with the unusually narrow width of the streets to give a sense of "wallness" to the French Quarter. There is no other city in North America with a street profile like this.

When I open the front door beneath the sunburst fanlight, I am almost knocked over in surprise by the vastness of the front hall. It must be fifteen feet wide. A majestic bronze chandelier hangs from the fourteen-foot-high ceiling, and the hall is appointed with statuesque candle holders and an oversized mirror. Straight ahead, an elegant curving stairway beckons me to explore the second floor, but my eye is drawn beyond it to a view of the sun-filled courtyard at the rear of the house. It is a centre hall designed to impress visitors—it certainly does the trick for me—and to welcome crowds of people for large parties.

I turn to my right and enter the grand parlour, a palatial room with heavy red drapery, a glittering crystal chandelier and a red, green, and gold-patterned carpet that spreads into the adjoining dining room. The focal point is a fireplace of black Egyptian marble. The furnishings include a sofa and chairs of curving lines and a pianoforte. It's an overtly opulent space in an opulent house, designed to impress visitors with the Hermann family wealth.

I love to tour houses. It doesn't matter if it's a small or a large place, plain or fancy, I find it a way to get to know people better, whether they're standing with me or have been gone for hundreds of years. I think about what it would be like for Franny and me to live in this house, to entertain in the parlour, to eat in the dining room and to cook in the kitchen. I try to get a feel for the place.

Over the fireplace is an 1832 painting of the Hermanns' daughter, Marie Virginie, at age sixteen. She is portrayed as a beautiful young woman with large brown eyes and brown hair coiffed in a bun, and wearing a blue silk dress. The Hermanns had three sons together, and Mrs. Hermann had two sons with her first husband. I can only imagine that, as the only daughter and the youngest child, the affections of the family and house would be focused on Marie Virginie.

I step through sliding doors and am seduced into the adjoining dining room by another grand crystal chandelier. Larger than the parlour, the dining room is anchored by a gleaming mahogany table that, with all its leaves in place, seats fourteen. I would die to have this in my house. I imagine the Hermann family sitting at the table for breakfast and enjoying their French toast. At the rear of the dining room is a door that leads to the china pantry from which dishes were served.

I leave the dining room and walk back to the front hall, where the curving staircase leads me up to a central hall on the second floor. Windows at both ends of the hall provide not just light but also air circulation, needed for a house in the humid South. Generous bedrooms are elegantly furnished with mahogany armoires, canopy beds with supple pillows, and wood stepstools to help crawl into bed.

But as much as I love exploring this amazing house, I want to get cooking again. If only I can find the kitchen. I take a narrow stairway, tucked away in a back corner of the second floor. Like many mansions, a stairway like this was used by staff so they could move about the house without being seen by family and guests. When I reach the bottom, I'm at the rear of the house.

Through the arched glass doors of the gallery I have a view of the three-storey brick kitchen structure, about twenty feet away. I push open the door of the loggia and step back into the courtyard.

Back in the kitchen I find Jennifer preparing the potager to cook the main meal of the day. With a trowel, she transfers embers and coals from the open fireplace to the lower holes in the potager. With the fire roaring in the fireplace and the potager heating up, the kitchen temperature continues to rise. Smoke fills the room and wafts out into the courtyard. It is not long before clouds of grey smoke fill the courtyard and rise above the foliage of the citrus trees. As the fire builds, I step outside for some fresh air. My hair, skin and clothes smell of smoke. I can only imagine what it would do to my lungs if I had to work in that room for the rest of my life.

Sitting and gasping on a bench outside the kitchen, I begin to imagine the enslaved people's lives and how each day unfolded in the house. A typical day would consist of two meals, breakfast around ten and dinner around two or three, when Mr. Hermann came home from his office. If only the immediate family was dining, dinner consisted of six or seven courses. (However, if the

Hermanns were entertaining guests, dinner could feature as many as twelve courses, with as many as forty dishes served over three or four hours.) After a two-to-three-hour meal in the afternoon, supper was usually optional. It would consist of a selection of lighter offerings—cheeses, crackers, snacks—with coffee and tea.

Before dinner, servants would layer the dining room table with white cotton tablecloths. After each course, while the guests sat at the table, servants would remove all the plates and cutlery, remove the top tablecloth and reset the table for the next course.

As the lady of the house, Mrs. Hermann selected the menu, double-checked that the table was set properly and ensured that the guests were comfortable.

Good china was used in the dining room every day. A trusted enslaved woman washed these dishes. To break a plate was a serious business, as plates were sold in sets and could not be easily replaced. Dishes were stored in the china pantry adjacent to the dining room. The pantry (the Latin *panna* or bread room eventually becoming storage room) would also have stored staples such as sugar, flour, salt, rice and cornmeal, in addition to spices, olive oil and imported delicacies like chocolate and brandied fruit. The pantry was typically locked to guard against pilfering of valuable spices.

Once the prepared dishes in their pots had been carried across the courtyard, other servants delivered the pots through the back gallery to the china pantry off the dining room. There, yet more servants transferred the food to serving dishes and garnished it, typically with parsley or lemon slices. Then they carried the serving dishes into the dining room, arranged them on the table and removed the covers. One or two servants would remain in the dining room, standing off to the side ready to deal with any further requests during dinner.

A dinner-party menu might look something like this:

First Course

Soups

(Creole varieties included crab with onions, garlic and tomatoes; okra soup; gumbo; and a whole turtle out of the shell, boiled with beef, wine and spices.)

Second Course

Salads

(Popular in New Orleans were potato salad with a dressing of egg yolks, oil and parsley; cabbage slaw dressed with cider vinegar, egg yolks and butter; and asparagus salad with oil and vinegar.)

Third Course

Fish and seafood

(Typical dishes were boiled white fish layered with a white sauce of flour and milk; stewed redfish in a red wine sauce; and baked crabs stuffed with a mixture of crab meat, bread crumbs and butter.)

Fourth Course

Meat and vegetables

(Four to six main-course dishes might include braised pork with roast apples; duck with mushrooms and pistachios; and ham stuffed with truffles and pecans.)

Courses Five to Eleven

(Meat and vegetable dishes would be repeated a number of times with different combinations.)

Twelfth Course

Dessert

(At least four, such as marble cake made with molasses and cinnamon; cream puffs with a custard of cream, eggs and vanilla; and fried doughnuts seasoned with cinnamon, nutmeg and rose water.)

These lavish multi-course meals were served in the French style, in which all the dishes in each course were put on the table at once. But diners were not obliged to eat everything. They could choose from various dishes and leave others. Their privilege was that they had so much variety and could eat as much as they wanted.

I am amazed by the bygone eating habits in this grand house. It would have been like dining at an all-you-can-eat buffet every day.

Jennifer preps the main dish of scalloped shrimp, or shrimp étouffée, by gathering ingredients and bowls on the central table before deveining the shrimp with a small sharp knife. The sight makes me happy. I love shrimp. I step up to the table and volunteer to slice onions, about the only cooking skill I can provide in this 1831 kitchen.

I manage to not cut myself. Jennifer slides my sliced onions along with a spoon of butter into a pan and places it over the fire. The room fills with the hissing and aroma of frying onions.

Jennifer then prepares a roux, a thickener for sauces, by melting butter in a Dutch oven over the potager. She tells me one variation on the roux: using bacon instead of butter because it could be cooked to a deeper and darker flavouring. Jennifer uses a seagrass fan to stoke the embers in the potager to increase the heat. She slowly sprinkles spoonfuls of flour into the melted butter in the pot and stirs the mixture to make a paste. The roux bubbles and thickens as she blends it together with a wooden spoon.

Next, she adds my fried onions to the pot and then gradually stirs in milk to make a thick sauce. She asks me to pass her the bowl of raw shrimp, and I watch as she slowly folds the shrimp into the sauce. She sprinkles Worcestershire sauce, salt from the salt dish and pepper from a small bowl into the pot. A final touch is to add fried okra for an Afro-Caribbean flavouring. The Hermanns would have been served their étouffée over rice.

In French, *étouffée* translates as "smothered." Smothering is a Cajun and Creole technique in which meat or fish are gently cooked in a covered pan with a little liquid, similar to braising on the stovetop. Smothering was used to render tougher cuts of meat tender and flavourful. Today smothered dishes are a standard in Louisiana restaurants, served over boiled white rice.

Jennifer offers me a spoonful of the étouffée. It is truly divine, a dish that deserves praise today just as it did in 1830. There is definitely more smothering in my future.

My host returns to the table and slices mushrooms and more onions before preparing a sizzling mushroom ragout in a frying pan over the hearth fire. When the dish is almost complete, she picks up a tarnished metal box and grates a small ball of nutmeg over the top. In our meal today, the steaming and enticingly fragrant ragout stands in for the seven vegetable side dishes that would have accompanied the Hermanns' scalloped shrimp.

The Hermanns were not to enjoy their house and their luxurious lifestyle for long. They were casualties of the Panic of 1837, a recession in the United States that lasted into the mid-1840s. Profits, prices and wages went down, banks collapsed, businesses failed and thousands of workers lost their jobs.

As the English cotton market crashed, Samuel Hermann went bankrupt. By the late 1830s he was in substantial debt and in 1841 was forced to dissolve his assets and sell the house.

Because Louisiana laws allowed women to keep money from a previous marriage, Mrs. Hermann was able to use her own money to buy back the house for a time. Eventually, though, she had to sell it, and it went to a prominent scholar and judge, Felix Grima, in 1844.

We cannot leave the Hermann-Grima House without having my favourite dinner course. I always look forward to dessert, and if there is more than one, I am in heaven.

The word *dessert* comes from the French *desservir*, meaning "to remove what has been served." Until the 1600s, sweet and savoury dishes were often on the table at the same time. Sometimes they were mixed together in the same pot with meat (mincemeat) or fish (eels) or sugary sauces. However, in the seventeenth century, French chefs decided that sweet dishes should be served at the end of the meal, in a separate course. Dessert was popular with the wealthy—sugar was expensive, and so only they could afford to indulge. And serving lots of desserts was of course another way to show off one's wealth and impress friends. Nevertheless, the new dessert course was a matter of great debate across Europe. But with the invention of such dishes as tarts, éclairs and petit-fours, the practice caught on. *Très bien!* What would dinner be like without dessert?

Jennifer begins making ours by melting butter and brown sugar in a saucepan over the potager. She adds Louisiana pecans and a pinch of salt and swirls them around in the pool of melted butter and sugar to coat them. As she spreads out the candied pecans on a tray to cool, Jennifer reminds me that the pecans would be only one of six or seven desserts that would be offered at the dining room table.

The smell of pecans roasting in butter and sugar drives me wild. My mouth waters as I wait for the nuts to cool. Dark brownish red, the pecans shine in the light with a coating of crusty brown bits. Finally, Jennifer gives me a nod, and I reach out.

Buttery, salty, sweet—the burnt sugar is so tasty. I adore it even more than chocolate. The charred sweetness combined with the intense salt, butter and pecan nuttiness is a New Orleans jazz party on my tongue. I cannot get enough of the crunchy, chewy, dark syrup deepness and I reach for more. Whether it is 1831 or the present day, candied pecans in pies, cakes and pralines are the essence of New Orleans dessert.

I've been happily smothered in New Orleans, one of my favourite cities in the world for eating. We have to get a piece of that into our kitchen.

Hermann-Grima House

820 St. Louis Street, New Orleans, Louisiana

www.hgghh.org | (504) 274-0750

Dear Franny,

Here are some quick notes from the Hermann-Grima House for our perfect kitchen.

Mrs. Hermann continued the exterior stone patio right through to the kitchen floor to blur the division between inside and outside.

We can plant fragrant flowers and trees outside the kitchen in the same way that the Hermann-Grima courtyard is blessed with citrus trees. Open doors and windows can coax outside breezes and scents into the house and kitchen. I think back to Monticello and how the cooks could step out of the kitchen into Jefferson's garden for fresh air and to pick vegetables.

Safety first—Keep a fire extinguisher in the kitchen.

Dishwasher (and washer/dryer?) in the same quadrant as the kitchen so that all washing is in one area of the house.

For creative cooking we can embrace the spice of life. In 1830, the Hermann kitchen cooked with different spices from around the world. They were southern foodies. Let's go all out with spice.

Not to worry. We won't be entertaining three hundred people at a time. That would be a lot of egg rolls for me to make.

XOXO

J.

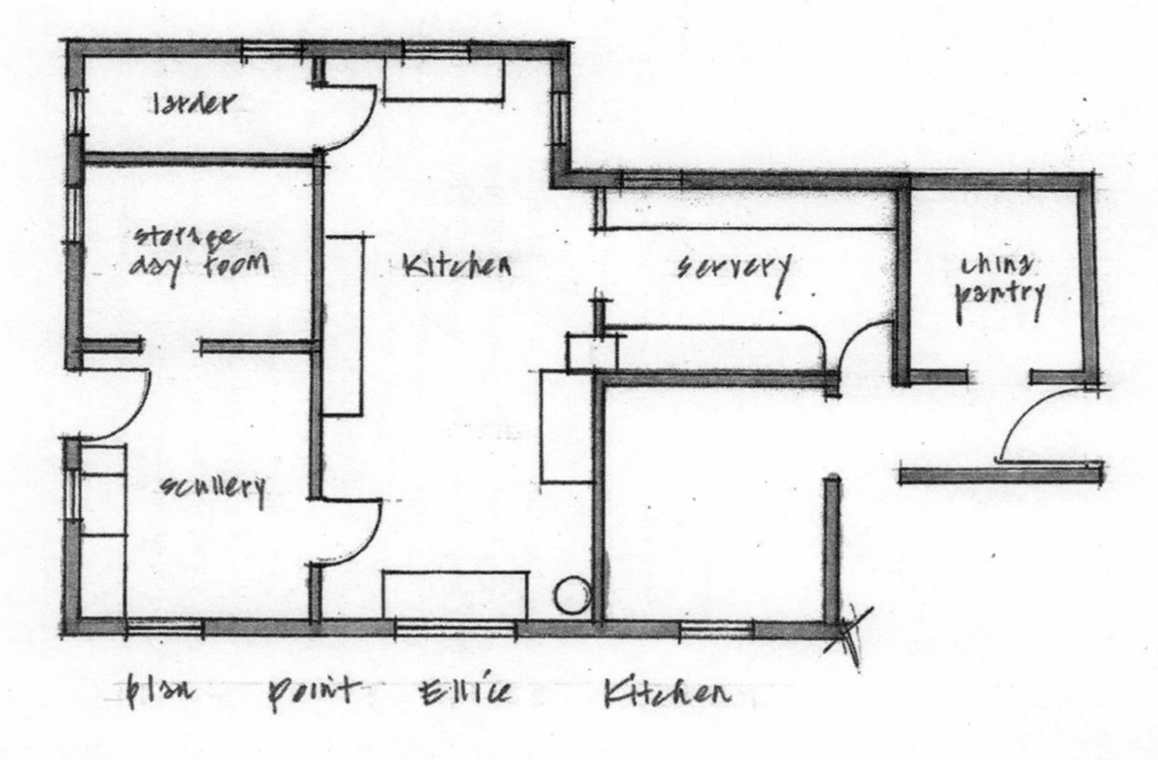
larder
storage
day room
Kitchen
servery
china
pantry
scullery
plan Point Ellice Kitchen

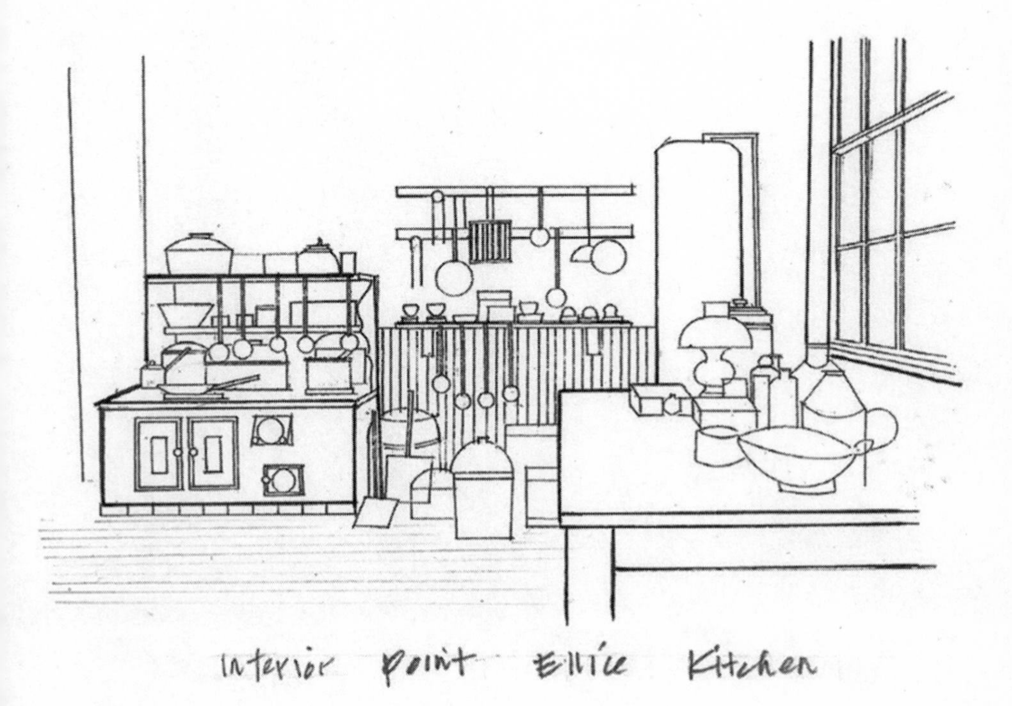

Interior Point Ellice Kitchen

4

POINT ELLICE HOUSE KITCHEN, 1890

Victoria, British Columbia

We all have a mental image of what it meant to be Victorian—the restrictive clothing, afternoon teas with cucumber sandwiches, games of croquet on the lawn and the fabulous country houses.

On my way to learn more about life in the late nineteenth century, I am strolling the gardens at Point Ellice House next to the Pacific Ocean in Victoria, British Columbia. On this journey, I want to explore the kitchens of the time from 1837 to 1901, when Queen Victoria reigned.

Almost hidden behind the lilies and rhododendrons is the rambling fourteen-room house that is restored to showcase the lives of the O'Reilly family in 1890. With its low-pitched roofs, overhanging eaves and rows of ornate brackets, Point Ellice House is a perfect example of a Victorian Italianate dwelling.

My Japanese grandfather started life in North America as a houseboy in a Victorian house in turn-of-the-nineteenth-century Chicago. I never met him, but I would guess that he spent a lot of time in the kitchen. Like the Pilgrims at Plimoth, his story is the same as that of so many other immigrants seeking a better life in a new land.

As a houseboy, he saved his money with a goal to attend the University of Chicago medical school. When his employer discovered this, he contributed money for my grandfather's education and made sure that his work schedule did not interfere with his studies. My grandfather graduated and moved to Vancouver, where he had a family, opened a medical clinic and treated hundreds of people.

Our family owes that Chicagoan a lot. I think one way to show gratitude is to be good to other people, and when somebody with potential comes along, to help that person succeed.

I did some research and found that the O'Reillys had a Chinese houseboy. Maybe this is why my first thoughts are of the servants who would be cooking, cleaning and butchering in the kitchen, as well as listening for the ring of the servants bells calling them to attend to a member of the family. If I had been born during the Victorian period, that houseboy could have been me.

As I ponder my life as a Japanese houseboy, I am met on the hard-packed gravel path by Gail Simpson, president of the Point Ellice House Preservation Society. She welcomes me to the house with a big smile and tells me she will be my tour guide for the day.

"Point Ellice House has one of North America's largest collections of Victoriana," says Gail, telling me that the family left behind all their belongings, including silverware and kitchen implements, when they sold the house to the British Columbia government. "I know you're going to love it—it's packed with Victorian artifacts."

I confide to Gail that I am leery of Victorians, with their straightlaced reputation and unsmiling photographs. Gail reassures me that the O'Reillys were a lively family who loved to party right into the next morning. They had a passion for food and drink.

In 1890, Point Ellice House was a gathering spot for Victoria's social elite. Peter O'Reilly was a retired gold commissioner and judge for the colony. He was also an Indian reserve commissioner (and one who seems to have been less than fair to Natives in his land deals.) He and his wife, Caroline, bought Point Ellice House in 1867, and over the years the O'Reillys had four children.

The house was the site of frequent dinners, afternoon teas, lunches, and croquet, tennis and boating events, as well as all-night dinner and dancing parties. Sumptuous food prepared in the spacious kitchen at Point Ellice House was a much-anticipated feature of the parties, with ingredients sailing into the Pacific port of Victoria from around the world.

I can hardly wait to get into Point Ellice House and check out the kitchen. Whether they're from Victorian times or today, I love to "meet" people who like to eat and drink. Even if at one time in history, I would have been their houseboy.

"I know you're anxious to see the kitchen," Gail says, as if reading my mind. "And I hope you like curry."

"Curry?" I reply. "Yes, I love curry. But what has it got to do with a Victorian kitchen?" I am a little perplexed, but I detect a twinkle in Gail's eye.

Aha! I like surprises, especially when it comes to food.

We are standing in the Point Ellice landscape, which is an explosion of colourful flower beds, an extensive vegetable garden and a series of woodland walks.

The O'Reillys worked hard to transform the forest around their house. Enticing views are framed by arching tree canopies, curving paths disappear around bends, flowers are planted at turns to lure the eye. Victorians used their gardens as a place to relax and entertain themselves, but they were also expected to produce food. At Point Ellice, the family depended on the land to produce its fruit and vegetables. This echoes back to the Pilgrim and Monticello vegetable gardens that were also counted on to put food on the table.

We approach the house along a winding path. I get only partial glimpses of the architecture tucked behind trees and shrubs, but I can see that the house is a bungalow in the Picturesque style. Quaint and natural—like a picture. The ideal of the Picturesque movement was that the house would harmonize with the natural landscape. Colours and materials used on the house reflect those of soil, rocks, wood and bark. The exterior walls are painted a pale rose, with brown railings, brackets and ornate porch posts. Dark red accents highlight the edges of the bay windows and gables. A close-up look at the stucco walls reveals that they have been scored to give the appearance of stone.

We arrive at the front porch with its double glass doors and glass transom. Gail invites me to press the latch on the door, and as I open it, I feel like I've stepped back 125 years.

I gaze around at the opulent gold-and-umber wallpaper, elaborate woodwork and heavy draperies that hang from the ceiling. An enormous gilt mirror hangs on the wall, the floors are covered in oriental rugs, and small panes of glass in the front door have been hand-painted in orange, blue, green and yellow to resemble stained glass. I covet a croquet set that features mallets, balls and cast-iron hoops leaning against the wall. As a kid, I used to love playing croquet in our backyard.

The walls are meticulously painted to mimic marble, and my

every move is monitored by the glassy eyes of a huge mounted Irish elk head. A quick scout around reveals that the house is a maze of bedrooms, a washroom, a parlour and a study. Period tables, chairs and knick-knacks so personalize the house that it feels as though the O'Reillys have just stepped outside to finish their round of croquet.

We wander into the grandest room of the house—the drawing room. I tell Gail that I enjoy how the Victorians indulged in an excess of ornament in their interiors, to reflect the wealth and prestige of the owners. Point Ellice House is over the top. There are tchotchkes everywhere.

The drawing room is set up around a gold-and-white marble fireplace whose mantel is completely covered in bowls, clocks, vases and figurines. I stand in the middle of the room and am swallowed up by chairs, tables, plant stands, lamps, vases, dishes, cushions and doilies. A multicoloured oriental rug spreads across the floor and adds to the visual commotion. Even the gold-framed mirror over the mantel incorporates a series of mini shelves to showcase more figurines. I almost expect Miss Havisham to shuffle out and scold me for my dirty fingernails.

We cross the hall to the extravagantly furnished dining room, where the mahogany table is, as expected, covered with glittering plates, bowls, glasses, silverware, crystal, folded napkins in goblets and ornate place cards. Victorian dining tables required precision planning. The table is set with the O'Reilly family's 154-piece set of Minton King's Border china, with its distinctive white well and burgundy-and-gold floral border. The various forks are arranged for fish, main course and ice cream, and there are knives for butter, cheese, game and fruit. Even the place cards are in curly, delicate ceramic holders that Mrs. O'Reilly bought in London.

My eyes are drawn to the two silver candelabras that rise above the table and set the tone of elegance and theatricality in the room.

Can I convince Franny we should have something the same in our dining room? With a large Chinese dinner gong outside the door and a piano for spontaneous musical interludes, this dining room reflects the importance of food and entertaining to the O'Reilly family.

When I try to picture what dainty little pastries the O'Reillys would offer for dessert, I remind myself of the purpose of this trip—the kitchen.

I step out into the hall to look for the kitchen. I peer around and realize I have no idea where it might be.

I try an inconspicuous closed door. It opens to a hallway completely cut off from the rest of the house. I enter and to my right is a small room, the china pantry (I am reminded of the Hermann-Grima House china pantry). Here, the maid or houseboy setting the table had ready access to the sets of dishes used for every type of meal enjoyed by the O'Reillys and their guests. The Victorians loved having cutlery, utensils and tableware for every purpose and occasion, and I can see that was true at Point Ellice House. Tiered shelves on all four walls are packed with colourful china and glassware of every description.

At the end of the hall is another door leading to a second passageway. This second hall has a narrow counter along one wall and cabinets displaying dishes. This must be the servery, where staff plated meals on the narrow counter and put final touches on dishes before they were sent to the dining room.

I know that Victorian architects tended to isolate the kitchen in order to contain noise, smells and the risk of fire. But the path to this kitchen is unusually elaborate. There are two separate hallways (the second hallway is the servery) that act as buffers between the dining room and kitchen.

I notice six small bells attached to the wall in the servery. It is the bell-pull system, a network of wires and chains within ceilings and walls that rings bells to summon servants. Servants could identify where the summons was coming from by the tone of the bell. The O'Reillys, like all families of the period, liked their servants to be as invisible as possible.

I turn away from the pantry and see a large room at the end of the hall. Gail joins me as we continue our walk and I feel elated when, finally, I step into the Point Ellice kitchen.

As I stand in the middle of the twelve-by-eighteen-foot kitchen, the first thing that strikes me is the vast number of kitchen implements hanging on the walls—they are everywhere. I could be in a Victorian kitchen supply store, surrounded by brooms, brushes, ladles, spoons, cherry pitters, apple corers, rolling pins and pots of every size and shape. In one corner sits a Victorian water filter, a cylindrical ceramic container into which impure water is poured through a filter on top and then stored as drinking water in a reservoir below. I have never seen so many utensils in one room. It is the opposite of the spare kitchens at Monticello and the Hermann-Grima House.

In grander nineteenth-century houses, the kitchen would be housed in a building detached from the main house, a short distance from the dining room. In urban Victorian houses, though, space was scarce and so the kitchen was typically squeezed into the basement. The houseboy and other servants would toil in an airless, smoky and gloomy space. But the kitchen at Point Ellice House is located at the back of the house owing to the low ceiling height in the basement. As I look around, I think that the staff was fortunate to have the kitchen on the ground level. Four large windows bring daylight and fresh air into the room.

The kitchen is functional and without ornamentation (if you don't count all those utensils). A table in the middle of the room is the primary work surface. The outer edges of the kitchen hold the cooking stove, and a dresser and shelves that contain canisters of dry goods and spices. The only visual embellishment is wainscoting painted yellow to match the peeling umber wallpaper. The floor is covered in simple maple floorboards.

The Point Ellice kitchen is equipped with free-standing, movable furniture, as built-in kitchen cabinets had still not been invented. No wonder the utensils hang on the walls—there was no other place to put them! The only enclosed storage is provided by a dresser, similar to what you might find in a bedroom. The dresser is built with cupboards and drawers under a broad counter—the only counter space in the kitchen. Above are two shelves, running the length of the dresser, on which jugs, moulds and various utensils are kept. Larger utensils are stored in an open space under the counter.

The stove's intimidating presence reminds me of the Dowager Countess of Grantham in Downton Abbey, *dressed in black, giving threatening looks.*

"The salmon poacher was a key piece of equipment here," says Gail as she points to the elongated metal pan on a shelf over the stove. "Salmon dinners were a favourite of the O'Reilly family." Point Ellice House is, after all, in the great Pacific Northwest, the land of the noble salmon.

Six feet across and three feet deep, the black cast-iron stove dominates the spacious room. The range is covered in pots, pans and kettles. Ladles and spoons hang off the two shelves on the wall above it. The stove's intimidating presence reminds me of the dowager countess of Grantham in *Downton Abbey*, dressed in black, giving threatening looks. The words *French range* are embedded in the cast-iron face, but it might as well say, "Don't Mess with Me."

Used by the family until the 1960s, the coal- and wood-burning range was made in Victoria at the Albion Iron Works. The stove was floated to the house on a barge, and it took ten men to carry it up to the kitchen. I can guess that once the fire inside was roaring, the cook would have to be a nimble dancer, opening and closing the mysterious doors and handles as she cooked for the family. Beside the stove is a box of coal to fuel the fire. In the corner of the kitchen is a copper hot water tank, heated by tubes running through the stove.

By the 1850s stoves had become common in the kitchens of middle-class homes. And it was all for the better.

Presenting the top 5 reasons why the stove was a cook's new best friend:

#5: Stoves were easier to use. No more bending, lifting and hanging heavy iron pots over an open-hearth fire.
#4: Food was less likely to burn. Temperature could be controlled with dampers.
#3: Stoves were safer. A stove did not spit out sparks and set the house on fire.
#2: Stoves were more efficient. Heat from the fire could be spread across the surface, meaning several pots could be used at the same time.
And the #1 reason: There was less risk of a cook's clothes catching fire—a hazard that for ages had led to serious injury or even death.

But stoves did not eliminate all the drudgery of cooking. The fire had to be started each morning and tended to all day, requiring about fifty pounds of coal every day. Ashes needed to be emptied out at least once a day. The dirtiest and probably most despised job in the home was cleaning the stove and applying heat-resistant wax to prevent rust. I am sure that the houseboy would be the one spending an hour or more each day cleaning the stove.

Gail guides me into the larder, a room used before the coming of the ice box for storing such produce as milk, cheese, butter and meats (French *lardon* for bacon—a place to keep meat). I have never been in one before, so this is a first. In some Victorian houses this room was used to store bread and pastry and was called a pantry. To keep it as cool as possible, the larder at Point Ellice House is situated on the north side of the house, to receive the least amount of sun. A window opposite the door coaxes in drafts of cool air. In some houses, the roof of the larder would be covered in wet towels on hot summer days.

When the servants of Point Ellice House were not cooking and cleaning up after meals, they were kept busy preserving food—by drying, curing, pickling and salting—making jams and jellies. In those days before refrigeration, this was not a matter of convenience, but a necessity, especially in winter. I realize that 170 years later, things had not changed much from the Pilgrim kitchen in Plimoth.

We next go into the scullery, a separate area at the rear of the kitchen, where the wet and dirty work was done: washing dishes, cleaning fish, plucking fowl, scrubbing and peeling vegetables. The Point Ellice House kitchen, like many in the Victorian period, has no sink. The kitchen was used only for cooking. All the messy, low-status activities that involved water were done in the scullery.

The scullery sink and drainboard are mounted against the wall, with a wood plate rack above it for drying dishes. The sink is quite shallow, to facilitate cleaning fish and vegetables. At the rear of the scullery is a screen-door entrance for servants and grocery deliveries. Though the Chinese houseboy did not live in the Point Ellice House, he was provided with a small day room beside the scullery—a tip-off that he handled the most labour intensive-cleaning and was the lowest-ranking servant.

The O'Reillys employed a cook, a maid, a cleaner and the Chinese houseboy. The houseboy lived in Chinatown. He started his day at five thirty or six every morning, lighting the fires in the house. Then he helped with chores such as cooking, ironing, washing, cleaning the house, grocery shopping, walking the family dog and taking care of children.

Immigrants from China in the nineteenth century were often peasant farmers escaping poverty. Most, like any other immigrants of the time, intended to work hard—as servants, cooking and cleaning—and return with their new wealth. Another common practice was to send money home to support the family. It was a hard life.

The white settlers considered the Chinese racially inferior and too visually and culturally different ever to become part of North American society. Yet here on the door of the Point Ellice House scullery I spot a photograph of the houseboy smiling and holding the hand of an O'Reilly baby. His hair is shaved in the front and in a long braid at the back. He is wearing a traditional Chinese tunic. In the Point Ellice archives there is a letter written by one of the O'Reilly sons to his father describing how the servant took him to Chinatown to watch the fireworks.

In the photo the houseboy looks happy and healthy. I hope he was. In a different time, I could have been him.

Life in many elite homes in Victoria was defined by strict rules of dos and don'ts, all dictated and enforced by the upper classes. The rules helped keep the status quo—the servants below and the wealthy on top. One of those rules was that the work of the kitchen was done by domestic servants (and the Chinese houseboy, if there was one). The lady of the house almost never entered. Typically, she would meet with the cook in the servery or the butler's pantry to plan the menu. Beyond that, the cook would be responsible.

Caroline O'Reilly, though, was an exception. Cooking was her passion, and her specialty was curry.

Mrs. O'Reilly shared cooking duties with her cook in order to pass on her skills, but sometimes she put herself in sole charge of preparing food, with the staff there for assistance and cleanup.

As members of the Victoria elite, the O'Reillys entertained notable guests, including Sir John A. Macdonald, Canada's first prime minister. Another dinner guest was Capt. Robert Falcon Scott, the Royal Navy officer and explorer. I wish I could have been at the table when Scott of the Antarctic dined at Point Ellice House. But I would not have been at the table. I would have been working in the scullery.

By the second half of the nineteenth century, curry and other Indian dishes had become fashionable among the elite and upper middle classes in England. Most of India was under British rule, and Queen Victoria carried the title Empress of India. Although the Queen never visited India, she was fascinated with all things Indian. She collected Indian paintings and employed Indian servants, including two cooks who regularly served curry at her dining table. Among the servants was Abdul Karim, with whom she had a scandalously close relationship for more than a decade.

Caroline O'Reilly learned South Asian cooking while visiting India. At the time, foreign travel was mostly undertaken by gentlemen to finish off their education, but a number of ladies also travelled abroad. Travel to India became cheaper and easier with the advent of the railways. Women like Caroline would voyage to India to visit relatives, write about the country and its customs and sketch landscapes. Later, after Caroline came to live in Point Ellice House in 1867, she entertained with curry dinners that were the talk of the town. Invitations were highly prized.

The itinerary for a typical party evening would be something like this:

8 to 9 p.m.: Drinks and starters
9 to 11 p.m.: Dinner
11 p.m. to midnight: Desserts
Midnight to 1 a.m.: Pipes and brandy for men in Peter O'Reilly's study. Tea and/or sherry for women in the drawing room.
1 to 4 a.m. or later: Dancing
4 a.m. or later: Guests leave for home
5 to 7 a.m.: Sleep
7 a.m.: Breakfast

The O'Reillys loved to party in the drawing room, and on those evenings, furniture would be moved to the side and the carpets rolled up for dancing with friends and other guests. Alcohol was allowed, and favourite potables included sherry, port, beer, and Scotch and Californian whiskey. Guests frequently stayed all night at the house.

Gail asks me if I am ready to make a nineteenth-century Victorian curry following the same recipe Mrs. O'Reilly used in 1890.

"Are you kidding?" I say. "I've had curry on my mind all morning!"

Gail guides me to the contemporary kitchen that is in a separate building next to Point Ellice House.

Before we cook, Gail opens up a dusty box. Inside are three cookbooks used by Mrs. O'Reilly in the nineteenth century. I know it's ridiculous, but I feel like I should ask Mrs. O'Reilly's permission to open these books—they are so personal. With their tattered pages and broken spines, the books are held together with ribbons. I put on plastic gloves and leaf through the pages. The recipe we are using today is an O'Reilly favourite called Hindoostanee Curry from the *Model Cookery and Housekeeping Book* by Mary Jewry, published in1868.

In the kitchen, Gail introduces me to my curry-cooking instructor for the day, a young woman named Nidhi Sheth. Nidhi, who is originally from Gujarat province, in the northwest of India, operates an Indian cooking school in Victoria and is an expert in Gujarati cuisine.

"There is not one way to make curry," Nidhi tells me. "The dishes vary according to each region—some are hotter, some are sweeter." While she talks she busily lays out her jars of spices, metal tins, chicken thighs and mortar and pestle on the central kitchen counter. "Indian women love to cook," she says in her high, melodic voice. "We are always in the kitchen. That is our occupation. We are also feminists, but we love to cook."

Nidhi brings a pot of water to a rolling boil over the stove and adds basmati rice.

I notice that in the nineteenth-century cookbook we're using, the cooking time for rice is given as four hours. "Rice was different in those days," Nidhi explains. "It was sold unpolished and with some of the husk remaining."

While Nidhi is making the rice, Gail and I chop onions and garlic for the curry. Although my technique is awkward, I get the job done. I pass the chopped onions and garlic over to Nidhi, who slowly fries them in oil until they turn soft and translucent. Mrs. O'Reilly's onions would have taken much longer to cook, working over her Albion Iron Works coal stove.

To prepare the chicken, we cut boneless, skinless thighs into bite-size pieces. "Thighs are standard for curries," says Nidhi. "They are flavourful and moist and can sustain a long cooking time."

Our next step is to boil the chicken chunks in water, into which Nidhi also drops cloves, mace, a cinnamon stick and a cardamom pod. She asks me to sniff each spice before she adds it to the pot. I revel in the pungent fragrances. This cooking class is also becoming aromatherapy. Nidhi removes the chicken from the pot after ten minutes of simmering and puts the broth to one side.

"Now comes the fun part," she says. She reaches for her jars of spices and herbs to make the all-important curry powder. I picture Mrs. O'Reilly doing the same thing at the table in her cluttered Victorian kitchen.

Spices were once a prized possession. The search for spices led to the discovery of the New World. Wars were fought over spices. Sailors were paid in pepper. Spices were what oil is today.

As in Mrs. O'Reilly's day, we'll be using a wood mortar and pestle to pound the spices. I have a flashback to my mortar and pestle in the Pilgrim kitchen. I remember how hard I had to work just to grind down a small amount of spice powder.

Into the mortar bowl Nidhi spoons turmeric powder, coriander seeds, poppy seeds, dried ginger, bright red Kashmiri chili powder and cumin seeds. I take the pestle club and start to grind away in a circular motion. After three minutes I check my progress and find that things have been brought to rough, seedy granules—the shells of the coriander seeds are still visible.

Nidhi looks into the bowl and scrunches her nose. "This is not good enough," she says. She takes the pestle from me and leans over the bowl. With an arched back and a low centre of gravity, Nidhi pushes all her weight onto the pestle and starts an intense back-and-forth grinding motion. I notice that her technique allows her to put much more force into the bowl. Presto. After about one minute of Nidhi's grinding, the bowl is full of fine powder. She's a human kitchen appliance. I draw close to the mortar and breathe in another wonderful dose of aromatherapy.

"Your spices have to be really fresh to make a great curry," says Nidhi. "Still full of their oils and flavours. Grinding and the heat it produces release the oils and their flavours. If the spices are stale, the oils and flavours have dried out." She invites me to smell store-bought prepared curry powder. It has nothing of the delightful depth and complexity of the freshly ground version.

Nidhi asks me, "Have you ever tasted these spices by themselves?"

"No," I reply. "I've never even thought of doing such a thing."

She pours a little bright yellow turmeric powder into my hand and tells me to taste it. I look at her a little perplexed, and then raise my hand and touch my tongue to the turmeric in my palm. My mouth fills with an unusual dry, warm, tangy flavour. My taste buds are alive. "So this is what turmeric tastes like," I say.

"Each spice has a particular flavour," says Nidhi. "Sweet, salty, pungent, spicy, sour, astringent."

We all know that each spice has a different flavour—that is their point. But how many of us take the trouble to become familiar with the individual characteristics of spices that, perhaps especially in Indian cuisine, we are most likely to come across in combination with others?

Next Nidhi puts ground coriander into my palm. Much more enthusiastic now, I lick it up. A herbal, lemony taste slightly puckers my mouth. I am amazed.

"If you don't know the taste of the spice," Nidhi says, "how can you know what it will taste like in the food?"

We are in the middle of cooking now, and Nidhi is scooting about the kitchen, chopping, stirring, tasting, all the time talking to me about spices, curry, Gujarat, cooking times and chicken thighs. It is all highly enjoyable and I am loving being part of the Indian ballet. Gail has taken a seat and is observing. She toured me around all morning and I think she needs a rest from all my questions.

Nidhi stirs the chicken into the onion and garlic mixture along with some of the chicken broth. She expertly rolls the chicken pieces with onion and garlic into our ground spices and adds some butter. I look longingly at the spiced chicken pieces in the pot. They've turned a bright yellow from the turmeric.

The culinary climax comes when Nidhi folds a few spoonfuls of coconut milk into the chicken. As the heavenly looking curry bubbles in the pot, the coconut milk lightens it to a creamy light yellow. When the gravy thickens, Nidhi turns off the heat and lets the curry stand for a few minutes. Waiting allows the spices to meld and mature in the pot.

While the curry is sitting, Nidhi busies herself setting out accompaniments such as anchovies, mango chutney, papadums, sliced cucumbers and bananas, and almonds.

After about ten minutes, Nidhi scoops the curry onto a platter with her hands. As she is scooping she explains, "We mix with our hands. We use our hands a lot in India."

We will be dining on the porch overlooking the garden. It is sunny outside, and in true Victorian fashion, the beauty of the house and garden come together and become one place. Gail sets the table with a festive Indian-print tablecloth. Nidhi and I plate the dishes, and we all sit in the spot where Mrs. O'Reilly would have entertained her nineteenth-century guests.

First, I scoop the hot and fluffy basmati rice onto plates. I pass them over to Nidhi, who spoons generous amounts of the gorgeous yellow curry over the rice. The steam of the rice and the aroma of the spices fill my nostrils. I am starving.

Finally, it is time to taste the curry. I load my fork with rice and a chunk of chicken smothered in yellow curry sauce. As soon as the curry touches my tongue, I feel a burst of fireworks. Immediately delectable and complex, the creamy, spice-filled sauce fills my mouth with a wild array of flavours.

"This is fantastic!" I say. There are so many flavours, I am happily confused about how and why they can work together so well. After four forkfuls in rapid succession, I ask Nidhi, "What am I tasting?!"

She looks back at me and replies, "You tell me."

I take another forkful and press my tongue to the top of my mouth. "I taste heat from the Kashmiri chili," I say. "The cooling from the coconut milk. There is dryness from the turmeric and pungency from the coriander." But when I concentrate and pull apart all the spices . . . "I taste the coriander the most, and first. A split second later all the flavours come together."

If I hadn't had the experience of tasting the coriander by itself, I would not even know what I was tasting. In this choir of curry spices, coriander has the most vibrant voice.

The curry is not overly hot. I can detect the chili but it is not overpowering, as in some curries I have had, and so its heat does not cloak the tastes of the other wonderful spices. That would be a culinary crime. The chicken is tender in my mouth, lightly spiked with clove and cinnamon. The spices in the sauce are deliciously moderated by the creamy goodness of the coconut milk. The side dishes of cucumber and almonds give contrasting texture and tastes that clear the palate. My taste buds are alive.

Nidhi and Gail pass around smaller plates of garnishes—the banana, anchovies and a jar of bright red chutney that glimmers like a ruby in the sunlight. The table offers a wealth of vibrant colours, aromas and textures. I think that if I were a nineteenth-century Victorian, I would have loved being introduced to this Indian cuisine.

Nidhi offers me a papadum. Crisp, paper-thin and almost

translucent, the round cracker is made of fried bean flour and crunches as I bite into it. I am hooked by the touch of oil, saltiness, garlic and spice.

But the crowning glory is Nidhi's mother's mango chutney. Sweet, sour, tart, tangy all at the same time, the chutney enhances the flavours of the curry. Nidhi tells us that her mother in India collects sour green mangoes in early March, shreds them and dries them in the sun sprinkled with turmeric. She then places the dried spiced mango shreds in a sugar syrup to cure.

"Please tell your mother that this chutney is a sweet-and-sour singsong in my mouth," I say.

I have a much greater appreciation of curry after being part of the kitchen experience and tasting the individual spices. I am staggered at the depth of flavour that Nidhi can draw out of her ingredients. It is a dinner fit for an Indian prince.

"Cooking and eating should be a social and cultural event," says Nidhi. "The food must smell good, look good, taste good, feel good. It is not just there to fill your stomach. It should be revered."

I lower my head and quietly say, "Namaste."

As we sit on the porch finishing our intoxicating curry dinner, the conversation turns to Mrs. O'Reilly and how she started making curries. In her early twenties, Caroline sailed from England to India to marry a young soldier. She was often seasick during the five-month voyage. When she finally arrived in India, she was told her fiancé had been killed in action. There had been no way to notify her while the ship was en route.

The captain of the ship offered to take Caroline back home, but she decided to stay in India for a short time. She asked servants to teach her about Indian spices and cooking techniques as a way to help her through her grief.

Later, in her thirties, Caroline married Peter O'Reilly. According to the affectionate letters that they mailed to each other while Peter was travelling in British Columbia, it was a happy marriage. Over the years, Mrs. O'Reilly used her battalion of spices from India for her stylish curry dinners. Nobody else in Victoria knew how to prepare Indian, and she did the cooking herself with the Chinese houseboy by her side.

An aberration of unbending Victorian rules, Mrs. O'Reilly boiled rice, sautéed chicken pieces and ground chilies and coriander seeds with the servants. Like many of us today, she was happy in the kitchen.

Huzzah.

Point Ellice House

2616 Pleasant Street, Victoria, British Columbia

pointellicehouse.com | ellicehouse@gmail.com | (250) 380-6506

Dear Franny,

It was great to visit Point Ellice House and see a Victorian kitchen up close. It was full of period knick-knacks. Here are some ideas that we can think about for our kitchen.

The spectacularly cluttered kitchen would not be appropriate for our house, but in an amusing way, I like it. I'd like to capture that creative and fun spirit in our kitchen.

Victorians situated the larder and kitchen on the north side of the house. The south side gets direct sun exposure, so it is the warmest. It makes good energy sense to have cooking and food storage on the cool north side, and dining on the south side. The O'Reillys were passive solar planning in the nineteenth century!

They didn't have much counter space, but the work areas were all at different levels, including the central work table, the dresser and the scullery counter. We should have different height levels for different kitchen tasks for personal ergonomics.

I like the open dish rack in the scullery. Imagine a beautiful wood rack mounted over the sink that allows dishes to drip-dry. I bet somebody is designing that right now.

To decorate our kitchen we can use wallpaper, paint, stucco, wood panel, marble, faux stone, gold flock . . . the Victorians offer a cornucopia of decorating textures and layers for the kitchen. The possibilities are endless. I dare you!

See you in the kitchen.

XOXO

Your personal houseboy,

J.

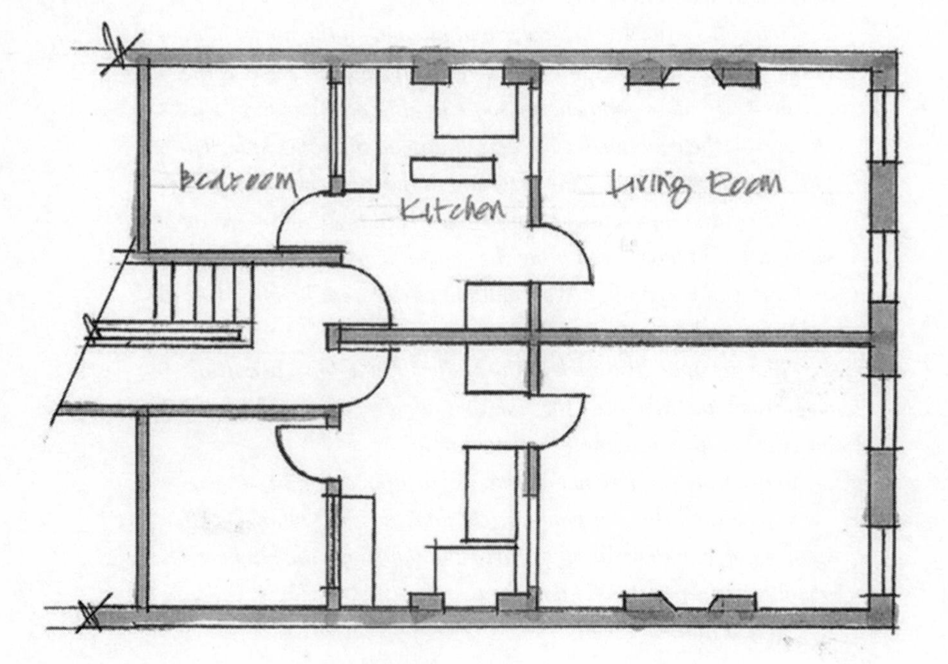

Plan Tenement Kitchen

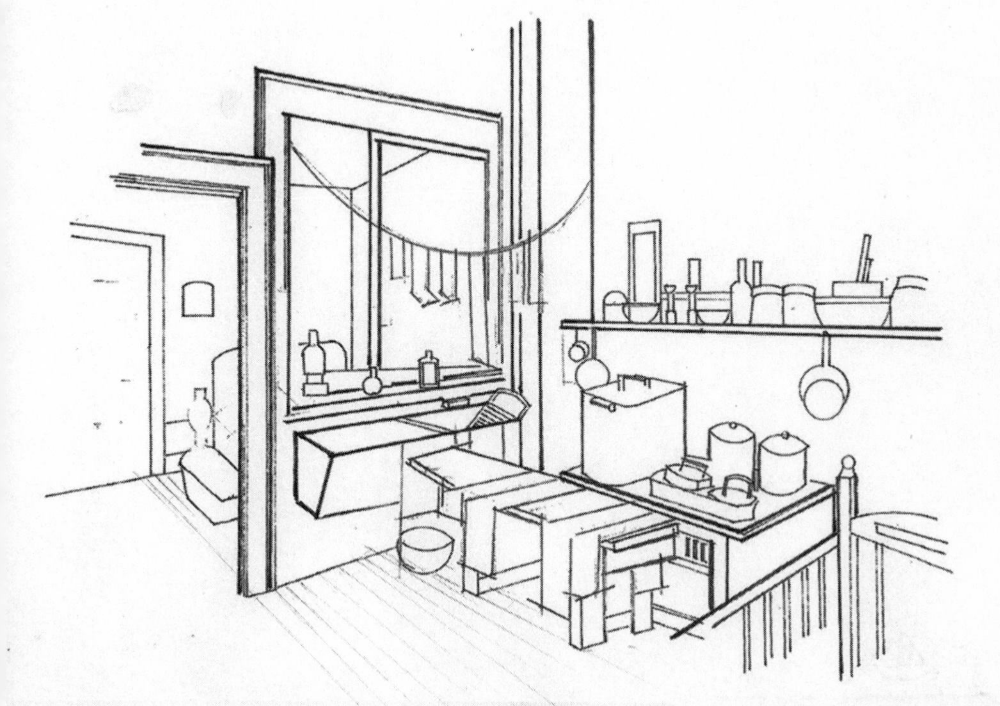

Interior Tenement Kitchen

5

LEVINE TENEMENT KITCHEN, 1897

Lower East Side, New York City, New York

Art, architecture, theatres, museums, galleries, music, shopping . . . everything in New York is exciting and larger than life.

I always love to walk the streets of New York. But I am feeling even more psyched on this trip because I am with my former architecture-school classmate Annie Lewison. Annie has lived on the Lower East Side for many years and she knows every building, street, back alley and crack in the sidewalk in her 'hood.

During her time here, Annie has brought up two great boys as a single mom, shopped, schlepped and run a household. If that wasn't enough, she has also worked as an architect on projects such as the Holocaust Memorial Museum by I. M. Pei, the World Trade Center Transportation Hub for Santiago Calatrava and the National September 11 Memorial and Museum by Snøhetta Architects. "New York was supposed to be a stop-off on my way to Israel," Annie tells me. "But thirty years later, I'm still here."

When Annie told me about a place in New York called the Tenement Museum, I knew I had to visit. It comprises, she explained, two well-preserved time-capsule tenement buildings in which are recreated the apartments and businesses of nineteenth-century immigrant families. The kicker for me, of course, is that the tenement kitchens are still intact.

It is just before nine in the morning and Annie is escorting me to Orchard Street, site of the Tenement Museum, on her way to work. We are in the morning rush and everything seems to be moving in triple-fast motion. As I cross the street I have to do a two-step around honking yellow cabs, buses coming at me like big circus elephants and bike couriers bouncing in and out of traffic like pinballs.

I am especially interested in seeing the Levine apartment at the Tenement Museum. I've read that the Levines arrived in the Lower East Side from Poland in 1890 and used their apartment not only as a place to live but also as a small garment factory to support themselves and their five children.

The Levines' apartment, like many on the Lower East Side, was a sweatshop where renters juggled home and work life. I am prepared to see the worst.

Amid the cacophony, Annie and I maintain our stride. We turn onto Orchard Street, a deep and narrow canyon between five-storey brick buildings that cast deep shadows on either side. I peer into windows of boutique hotels, cool bars and swishy clothing stores that are shoehorned between century-old hat merchants, hardware stores and clothing shops. It is a neighbourhood in transition.

It is hard to believe, but in the 1830s, the Lower East Side was a neighbourhood of single-family row houses. But by the 1850s, thousands of new immigrants from Europe were arriving in New York looking for a place to live.

On Orchard Street, landowners knocked down their row houses. On the lots where once had stood single buildings, new ones were built that could hold twenty or more families on five floors. Each tenement—named for the tenant renters—occupied nearly the entire twenty-five-by-hundred-foot lot. Only the rooms facing the street received any light. The interior rooms had no ventilation, and there was no indoor plumbing. Thousands of these tenements were built on the Lower East Side in the nineteenth century.

By the late nineteenth century, Eastern European Jews, fleeing pogroms at home, became the dominant ethnic group on the Lower East Side. Annie's own grandparents escaped the anti-Semitic mobs of early twentieth-century Russia to make a better life for themselves in North America.

Annie and I have both worked on historical building projects, and a great treat for us is scanning a facade in detail for the first time. We stand back and take in the entrances of the tenement buildings, each consisting of steel stairs that rise up to the front door. The upper four levels have bands of arched windows and a steel fire escape, an iconic image of New York tenements. We happily take note of some graceful Italianate details that were in fashion in the 1860s, including brownstone lintels and sills around the windows. The tenements are sparse but handsome.

"I'm glad the cornices are in such good shape," I say to Annie as we look up and see that the buildings are topped off with ornate cornices and brackets made of cast metal.

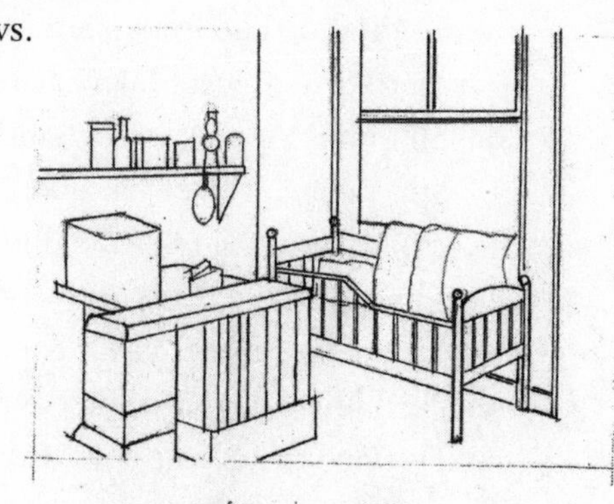

A cornice—a favourite feature of mine—is a ledge at the top of a building that redirects water away from the walls, preventing damage. They can also be wonderfully decorative. The cornice, which originated in classical Greek architecture, adds so much visual interest to North American residential and commercial architecture. Nothing hurts my sensibilities more than seeing a historic building with a cornice that has gone AWOL. I get great pleasure from looking up from sidewalks to see elaborate brackets, mouldings and dentils.

"This is it, Ota," says Annie as we reach our destination and stand in front of 97 Orchard Street. I gaze up. It is easy to overlook, because it blends in with the other tenements that form a continuous building wall along the street.

Annie needs to get to her office. After work, she says, we'll go back to her place to make a very special dish. "OK, Ota, have a good day, but be hungry tonight," she says and leaves me at the stoop.

At the front door I am met by Miriam Bader, director of education for the Tenement Museum. We stand at the top of the stairs and imagine the Orchard Street neighbourhood at the end of the nineteenth century. According to some reports, the Lower East Side was one of the most densely populated places in the world.

"In 1897, this street would be a lot busier than it is today," says Miriam. "The front stairs, as well as neighbouring stairs on Orchard Street, would have been alive with children playing and neighbours stopping to chat and socialize."

Miriam opens the front door and we step into a dark, cramped space. My eyes adjust to the dimness and I begin to make out a vestibule with tile floors and scuffed wainscotting. I look up and I see crumbling plaster walls, panels of peeling paint and a ceiling of cracked pressed metal. Directly ahead of me is an old wood stairway and beyond that a rear door with a window, the only source of light.

I'm shocked. I feel like I've stepped into a derelict building that is about to collapse. Nothing has been fixed for years. And on top of everything else, threads of burlap hang from the old ceiling. It looks more like a haunted house than a museum. This is the exact opposite of the pristine historical houses that I am used to visiting.

But slowly I relax and I begin to get it. This is the real thing. The Tenement Museum is curated to give me a genuine experience of what life was like in 1897. It is not sanitized or romanticized. Instead, I am allowed to see, smell and walk through the same dark, cramped spaces as the residents did over a hundred years ago.

Miriam and I walk to the staircase that runs up the centre of the building. It has a heavy newel post and railing. With twenty families living in the building, I know that this was a busy staircase, with dozens of mothers, fathers, children and teenagers coming and going at all times of the day. I run my hand over the banister and try to picture the Levines dashing up and down carrying fabrics and dresses.

We go up to the second level and begin to walk along the dark and windowless corridor. In the gloom I find two cramped closets that open right onto the hallway. They're toilets. Flush toilets with wooden seats. My first reaction is that they offer little privacy. I've never before seen shared toilets in a residential building. The original 1863 building had no flush toilets, no showers, no baths and no running water in any of the apartments. Everybody had to use a privy at the back of the building. These flush toilets were finally installed in the hallways in 1905.

"They must have been a godsend," I say to Miriam, imagining tenants running down five flights of stairs for relief in the middle of the night.

Continuing down the hall, Miriam brings me to the door of the Levine apartment. I stop in front of the entrance door and take a deep breath. This is why I came to New York—to see a real nineteenth-century tenement kitchen.

I step into a dark apartment. The air smells stale, as if the place hasn't been ventilated for years. Slowly my eyes adjust, and I can barely make out two large windows about twelve feet away with the blinds pulled down.

Miriam opens the blinds to allow sunlight in, and I find myself in the living room of the Levine apartment. In the middle of the twelve-by-eleven-foot room is a dress form in a nearly completed pink and black gown. A sewing machine sits on a table nearby. The space is smartly finished with mint-green wallpaper and beige pull-down blinds with a decorative black fringe.

Clearly, we have found our way into the sweatshop. It may be messy with fabrics and threads, but it is not the blighted and inhumane hellhole of my imagination. It is clean and even rather cheerful.

Harris and Jennie Levine came to America from Russia in 1890 with few skills and little money. To provide for themselves and their five children, the Levines, like many immigrants, turned their living room and kitchen into a small factory, sewing together pieces of cloth supplied to them by clothing companies. Their sweatshop was one of more than a hundred in business on Orchard Street.

Harris operated the sewing machine while two employees worked alongside him in the living room. In the kitchen, a presser ironed the completed garments next to the stove that was used for cooking meals. Workers were paid a small amount for each completed piece and tended to work long hours.

We step out of the living room and back into darkness. Miriam flips a light switch and I find myself in the Levine kitchen.

It's windowless, sparse and tiny—about nine feet by nine feet. The walls are painted a cracked, bluish green. The floor is bare wood. The kitchen is straightforward, unadorned and functional.

A loaf of bread, dishes and a knife sit on a kitchen table that is squeezed into a corner. A wood shelf attached to the wall holds cups, bottles, pots and pans. On the floor are brushes and wash tubs for cleaning clothes. Socks and undergarments hang as if to dry on ropes across the length of the room. It must also have been a hive of activity.

Connected to the kitchen is another small space, an eight-by-six-foot bedroom completely shut off from both fresh air and natural light. It's the only bedroom in the apartment. Along with the living room and kitchen, the entire flat totals 325 square feet. All of the apartments in the building were the same size and layout, and it wasn't unusual for them to house families of eight or more. The Levine family eventually numbered seven, but at busy sewing times they would have an additional three employees sleeping in their apartment. In winter, the kitchen was the most popular place to sleep. It was the warmest room.

Dominated by a central heavy black cast-iron stove, this kitchen, in addition to its cooking role, was a workshop, a laundry room, a place to wash dishes and clothes, and a bedroom for visitors or employees who worked into the night. Jennie Levine would have shared the kitchen with the presser, who would have heated his heavy irons on the stove. I try to picture Jennie making a family meal, boiling diapers and washing clothes while dancing around the presser and the ironing board that was set up in front of the stove.

On the floor is a steel coal bucket. It would have weighed about forty pounds when full, a burden Jennie would have had to regularly carry upstairs from the basement coal bin in order to keep the stove burning.

Water came from an outdoor pump at the rear of the building by the row of toilets. Jennie would have had to fetch water from the pump and then, like the coal, haul it up the stairs by herself. Women like Jennie Levine were like Sherpas, lugging groceries, coal, and children up and down the tenement stairs. Just thinking about it makes my plantar fasciitis act up.

Food prepared and eaten in the kitchen fuelled the enterprising Levine family. Lower East Side immigrants favoured recipes from the old country. For Jennie Levine, these would have included gefilte fish, matzo ball soup, potato kugel, spaetzle, latkes and stuffed cabbage. When there was frying to be done, it was done in goose fat.

Sausages and frankfurters were considered un-American. Garlic was thought to make a person aggressive and nervous.

Cooking dishes from the homeland on their cast-iron stoves was one of the ways that immigrants could adjust to life in America. Homesickness was common, and familiar food, tastes and ingredients helped ease the emotional trauma of adjusting to a new world. It seems outrageous to me today, but immigrants were discouraged from cooking their native dishes by American community groups such as settlement houses, churches and the YMCA.

In her book *97 Orchard: An Edible History of Five Immigrant Families in One New York Tenement*, Jane Ziegelman, the director of the Tenement Museum's culinary centre, writes that the public school system in particular tried to Americanize the tastes of immigrant children and wean them away from pickles and deli meats in favour of oatmeal and boiled vegetables. Pungent foods such as sausages and frankfurters were considered un-American. Garlic was thought to make a person aggressive and nervous. Local schools sent dieticians into tenement kitchens to convert families away from strong-tasting food. Newcomers were offered cooking classes to learn to make "American" dishes such as pot roast, potato salad, cakes and cream pies. Mothers were told that unless they cooked blander foods, their children would not become Americans.

My goodness, what could be more American than a hot dog? I am crazy for hot dogs—especially at baseball games.

As my tour of the Levine apartment winds down, I gaze ruefully at the stove. "As much as I love to cook," I tell Miriam, "I'm not even going to try and make anything on *that*."

I take a last look around the matchbox-size apartment. "I'll never complain again."

I meet up with Annie outside the museum in the late afternoon and we head back to her apartment to cook dinner. On the way, she takes me on a Lower East Side food tour to sample some of the street foods from the late nineteenth century that can still be found in her neighbourhood.

During my visit to the Tenement Museum I learned that pickles were a popular food with tenement dwellers, who liked the sour garlic flavour. A common lunch among immigrants was a pickle with bread—a sandwich that would cost one or two pennies to make. Women like Jennie Levine would have made pickles for their families, and with no refrigeration, pickling was a practical and tasty way to preserve vegetables. Eventually the demand for pickles became large enough that tenement dwellers set up home businesses making pickles, and pickle vendors sprang up along Orchard Street and in the surrounding neighbourhood.

While at one time there were more than a hundred vendors selling pickles on the Lower East Side, today only one is left standing. We drop by a store called the Pickle Guys and Annie encourages me to try the pickled tomatoes. I am hesitant, never having heard of tomatoes being prepared that way. But when I bite into one, I immediately fall in love: small green tomatoes with big crunch and a subtle dill flavour. I buy a half-dozen plus a half-dozen dills, and we head back out to the street.

Annie shepherds me to Yonah Schimmel. It's the oldest knishery in America, and looks it. Nothing appears to have changed since it opened in 1910. A knish is an eastern European dough pocket filled with a beef or potato mixture. The recipe arrived in America with Jewish immigrants like Jennie Levine. I am treated to a pillow of puffy dough the size of a softball, filled with mashed potatoes, onions and mushrooms. It is extremely carb heavy, and I come away with a greater appreciation of the knishery's nineteenth-century tin ceiling, display cases and peeling paint than of the product on sale.

(Growing up, we had a Jewish woman living in our house who made us fabulous meat and kasha knishes. She also taught my mom how to make them—yes, a Japanese Canadian mom making vintage knishes. They were much lighter than those at Yonah's, with a thinner pastry and less dough. Just the thought of biting into the delicate, crispy pockets of thin potato crust pastry caressing a warm, fragrant mixture of ground beef, onions and kasha brings tears of remembered joy to my eyes.)

Delicatessens on the Lower East Side invented that classic New York sandwich, pastrami on rye. Like corned beef, pastrami was created as a way to preserve meat before there was refrigeration. Annie and I stop at Katz's Deli, a fun, popular restaurant, and pick up a huge sandwich filled with at least a dozen layers of pastrami. The aromatic meat has an intense smoky taste. But after one bite, Annie snatches the sandwich away from me so that I don't spoil my appetite for our upcoming dinner.

Back on the sidewalk and heading toward Annie's place, I tell her that we're all lucky that immigrant women held on to their recipes and didn't capitulate to making cream pies and rice puddings. I can't imagine America without hot dogs, pastrami and dill pickles.

—

I am standing in Annie's cheerful kitchen in her twelfth-floor apartment on the Lower East Side. Its walls are festooned with multicoloured mugs and dinner plates, postcards from around the world, pictures cut from architectural magazines, Thai shadow puppets; the room has a spirit that matches Annie's lively personality. I peek out the window and admire a splendid view of the New York skyline, with the Empire State Building, which seems to be centred on her kitchen table.

I think about the Levine residence just around the corner. Annie's kitchen has gleaming new appliances, but it is not dramatically larger than Mrs. Levine's. In New York, space has always been at a premium.

I am always happy to be in Annie's kitchen, but never more so than now. We are about to make the quintessential Lower East Side Jewish dish, matzo ball soup. Not many dishes have a history as sacred as this one. A chicken soup accented with dumplings, it's made every year in traditional Jewish homes for Passover.

"It's the ultimate budget meal," says Annie. "It's something substantial created from crumbs."

As Annie organizes the ingredients on the kitchen table, she tells me that the one ingredient that can't be bought at the grocery store is memories.

The secret to matzo ball soup, she explains, is "thinking about your parents, your family and all the people you shared this soup with every Passover. Memories of how sick you once were and how someone took care of you and how the soup made you feel better."

Wow, this is pretty deep. This is why matzo ball soup is so special to Jewish people. When I cook, I sometimes get impatient with all the shopping, cleaning and chopping involved. But making matzo ball soup seems to involve a cathartic, spiritual element.

The star of the show is the chicken. I have picked up a plump four-pound bird at the butcher, who assured me this baby will do the trick.

Annie, a chicken soup virtuoso, begins by instructing me to place the chicken in a soup pot and add enough cold water to just cover it. In a déjà vu moment, I picture myself back at the Pilgrim open fire pouring water into a pot to make braised duck—only this time there is no smoke in my eyes.

My next job is to chop vegetables—carrots, onion and celery—into rough chunks. Annie adds these to the pot along with pinches of salt and pepper and some sprigs of dill. I notice she doesn't use any measuring cups or measuring spoons.

"I enjoy the solace of making soup," she tells me. "I get a lot of satisfaction in making something from scratch."

We bring the pot to a boil and then lower the heat to a slow simmer. Making the stock is a test of patience, Annie says. It needs to simmer over a low heat for two and a half hours.

One of my jobs over the next couple of hours is to carefully skim off the chicken fat that rises to the surface. Annie has found me a thin-edged spoon for the purpose. The fat is called schmaltz and for centuries it was a staple ingredient in the Jewish kitchen. But in these health-conscious, fat-phobic times, cooks usually use vegetable or canola oil instead. As I continue to skim, I think of Jennie Levine doing the same in her tenement kitchen and storing the schmaltz for frying her latkes.

"It's much easier to skim the fat from refrigerated soup," says Annie. "You don't lose so much of the broth." In the 1890s, the fire escape in cold weather would do the trick.

With the chicken and vegetables simmering in the soup, Annie and I get a chance to chat about the history of matzo ball soup.

When Moses led them from slavery in Egypt, the Jews had no time to pack food. Instead of the leavened bread they were accustomed to, God commanded them to eat matzo, an unleavened dough of flour and water that was baked flat in the sun. In recognition of the exodus of their ancestors, Jewish people today refuse to eat leavened bread during Passover. The only bread that observant Jews eat at this time is matzo.

For generations, chicken soup has been called Jewish penicillin, and medical researchers in recent years have found evidence to support the claims of Jewish mothers. Chicken soup helps us deal with the common cold in all sorts of ways—as a decongestant, an anti-inflammatory and a source of hydration, nutrients and, of course, psychosomatic comfort. As Annie says, "If it makes you think of your mom, that's good too."

After about an hour of chat, laughter and schmaltz skimming, it's time for us to make the matzo balls.

I am completely ready for this challenge. Having never made any such thing before, I summon up all my powers of concentration.

"Making matzo balls is an art in itself," says Annie. To begin, she beats eggs in a bowl with a fork and adds water and a little vegetable oil. We could have used our freshly skimmed schmaltz instead of the oil, but I'd be in trouble with my doctor if he found out. Then Annie pours in matzo meal and asks me to thoroughly mix these ingredients with my hands to make a moist dough. The resulting dough goes in the refrigerator for an hour. My next task is to boil a pot of salted water for the matzo balls.

I take a peek at the dough in the fridge. It looks like a bowl of wet cement. It's hard to believe that this awful-looking mush will soon be transformed into white fluffiness.

Annie wets her hands under the tap and gently rolls a small lump of dough between her palms. I try to do the same. But I can't do anything with the dough. It falls apart in my hands. What am I doing wrong?

Then it hits me: I forgot to wet my hands.

I swoop over to the sink, get my palms suitably damp and try again. This time, things go much more smoothly. My first attempt is the correct size—like a Swedish meatball—but a little lopsided. It's more of a matzo gourd. I am ready to announce my accomplishment when I look over and see that Annie has steadily made a dozen in the time it's taken me to make one.

With the pot of water at a rolling boil, we drop in our matzo balls, turn the heat down to a simmer and put a lid on. Annie uses a separate pot because starch would murk up our clean-looking soup. This is a purely aesthetic consideration. I am a visual guy, and it makes complete sense to me. Looks count.

For the next thirty minutes, we are in a crucial period of simmering. The texture of matzo balls can turn out light or dense, depending on the recipe and the skill of the cook. Enthusiasts classify matzo balls as "floaters" or "sinkers." Which will ours be, I wonder anxiously.

At first the balls sink to the bottom of the pot. I think we've blown it. As we peer into our steaming pot, Annie tells me a tall story about Marilyn Monroe from the time she was married to Arthur Miller. After being served matzo ball soup three times in a row at the home of her Jewish in-laws, she supposedly said, "Tell me, Arthur, isn't there any other part of the matzo you can eat?"

After about thirty minutes, the matzo balls miraculously swell and float to the top. Success!

Now it's time to separate the stock from the bones and the stewing vegetables. We place a sieve over a second pot. With each of us holding one side of our soup pot, we tip the pot over and the soup spills into the sieve and drains perfectly into the second pot. The bones and cooked vegetables are caught in the sieve. Both pots are steaming.

Annie strips the meat from the chicken carcass and places it in the fridge. She discards the vegetables, as all their flavour has passed into the liquid. The soup is strained and the broth is clear. It is an ambrosia of chicken soup. Mission accomplished.

Finally, we sit down at the kitchen counter to enjoy the fruits of our labour. Annie places three matzo balls in each of our bowls and ladles in the steaming soup, a shimmering golden broth that is as clear as a spring morning. As a finishing touch she adds a sprig of dill and some slender carrot shavings as a garnish. All of a sudden, it looks like something out of *Gourmet* magazine.

I dip my spoon into the bowl and taste with delight. The flavour is deep, wholesome and bursting of chicken.

Next, I try a chunk of matzo ball. It is perfectly al dente, firm but still fluffy. A perfect in-between texture. The mixture of soup and matzo dumpling radiates warmth throughout my whole body. The world is good.

As we sit at the table and sip, Annie reflects.

"You know, Ota, this is not just about eating," she says. "Matzo ball soup causes so many feelings to come out in me. The taste and smell bring memories flooding back—family, home, being together. The soup is simple. It's nothing special. But it's the cure for everything—from bankruptcy to broken hearts."

I never had a Jewish mother to nurture me this way, but the glow I'm feeling is authentic. I've discovered a classic.

I imagine Jennie Levine making matzo ball soup over her coal-burning stove in her tenement kitchen on Orchard Street.

She and her husband never became millionaires, but they managed to produce enough dresses to afford to move to Brooklyn and raise five children there. Harris became an American citizen and continued to be a dressmaker until he died, in 1929. Jennie died in 1945. They are exemplars of the intense work ethic and optimistic attitude that new immigrants brought to America in the nineteenth and early twentieth centuries.

From New York to Florida, the descendants of the Levines are now scattered along the American East Coast. They continue to visit the Levine apartment at the Tenement Museum.

The Levines lived the American Dream. They came to America with nothing. They didn't speak the language. They worked hard, stitched together dresses, boiled soup in their tenement kitchen and made a better life for their children and their children's children.

I think they're heroes.

The Lower East Side Tenement Museum

97 Orchard Street, Manhattan, New York City, New York

www.tenement.org | 1 (877) 975-3786

Dear Franny,

I learned a lot in Jennie Levine's kitchen at the Tenement Museum. What I learned were the intangibles.

A perfect kitchen is not just about food. It's also about people. It's about remembering the family and friends we've cooked with and shared food with over the years. As Annie put it to me while we cooked . . . the one ingredient we can't get at the grocery store is memories.

I learned that sometimes it's about slowing down, the way that Annie made the matzo ball soup. Taking time to make things from scratch, with our hands. Making a special meal and to spend time with friends and family—or with just each other. And to be grateful for that time together.

I learned that I like the idea of being reminded of our grandparents while cooking. The Levines' tenement apartment made me think about my own grandparents who came to North America without money or knowing the language. All they had were hopes for a better life. They had dreams and the courage to follow them.

Let's look for some photos of our grandparents and hang them in the kitchen to inspire us.

Speaking of inspiration, recipes get handed down from generation to generation. They're part of a family continuum. We have some handwritten recipes from both our moms. We can frame those and hang them in the kitchen too.

And the most important thing I learned from Jennie Levine's tenement kitchen was: a kitchen doesn't have to be huge to be workable.

I hope you like matzo ball soup. We're going to be having a lot of it.

XOXO

J.

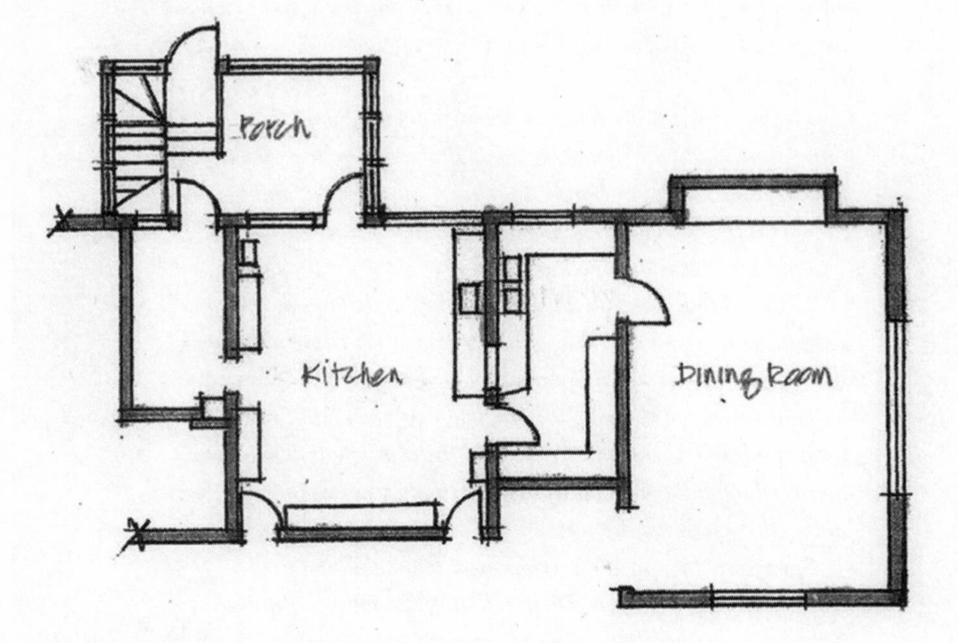

Plan Gamble House Kitchen

Interior Gamble House Kitchen

6

GAMBLE HOUSE KITCHEN, 1909
Pasadena, California

I am sitting at a picnic table overlooking the Japanese lily pond and clinker-brick terrace of a superb Arts and Crafts building in Pasadena, all the while spooning mouthfuls of walnut pie and whipped cream into my mouth. I have already partaken of too many glasses of California wine and am free of all of life's anxieties. Around the table are experts in California history and cuisine, and the conversation is lively.

Can life get any better? This must be a dream. Whatever happens, I beg you, please don't wake me.

I love my home city of Toronto. It does not have the monumental Champs-Elysées of Paris, nor the magical canals of Venice. But with a stable political scene and a burgeoning economy, it provides all the amenities of big-city life while balancing concerns for the disadvantaged, the environment and equality. I know of recent immigrants from war-torn countries who cry tears of happiness when they arrive in Toronto. Compared with other cities in the world, Toronto is an oasis.

But there's a catch: the Canadian winter. Even though there are colder parts of Canada, I find that the cold of Toronto is inhuman. Never-ending days of ice, snow and grey skies drive me to visit the depths of my human weaknesses—insecurity, pessimism and why am I so short? Before venturing outdoors, I don my long underwear, boots, coat, hat, gloves and scarf and lean into the wind. I suffer the February blues.

So it is no wonder that on a February morning, when I step out of a jet plane and a warm California breeze caresses my face, I feel that I am experiencing a new lightness of being. As my feet touch the tarmac at Los Angeles airport, I strip off my wool sweater and turn my face directly into the sun. Even though Los Angeles has a reputation for traffic jams, smog and quakes, today I have no concerns. In February, it is hard for me to hate sunshine.

In my quest for the perfect kitchen, I have come to L.A. to seek out the sublime woodwork and Zen aura of the iconic 1909 Gamble House, in Pasadena. Designed by the architect brothers Charles and Henry Greene, it is the ultimate Arts and Crafts masterpiece.

I want to see what natural wood in the hands of skilled craftspeople can do for a perfect kitchen. In our contemporary world of monochromatic white-on-white kitchens, I am still drawn to the warmth, smell and possibilities of wood. I also want to explore why the Arts and Crafts movement became so popular in Pasadena.

The Arts and Crafts movement is all about nature, wood and building by hand, and it is one of my favourite styles of domestic design. The movement began in England in the 1860s with the ideas of designer William Morris, who rejected mass-produced furniture of the Industrial Revolution in favour of natural beauty and traditional craftsmanship. The movement took off in America at the beginning of the twentieth century in reaction to the over-ornamentation of Victorian interiors. The palette comes from the beauty of nature: stones, bark, leaves, grasses. I especially like the furniture, with its natural wood graining, clean lines and exposed joints that showcase fine woodworking skills.

All my life I have been in awe of those who can cut, saw, plane and flawlessly put things together. I have built decks and ramps and framed additions, but not with the precision and ease of talented tradespeople. My complete admiration for anybody with superior carpentry skills only adds to my excitement at finally visiting this architectural masterpiece.

Before my visit, I drop in to see Becky Nicolaides, a good friend of mine who lives in Pasadena with her husband and two young children. She's a renowned urban historian, with a PhD from Columbia, and has agreed to be my guide.

Becky is also going to help me make an authentic 1909 Pasadena picnic lunch for special guests at the Gamble House. She tells me that as a historian who's specialized in the story of the L.A. suburbs, she's fascinated with the subject of picnics. I am so glad that we're in this together.

"In 1900, Los Angeles was up for grabs," says Becky as we drive up and down the hilly streets of Pasadena. "People came here to enjoy the warmth, sunshine, good health and California oranges. It was a new frontier. They were sick of putting up with the cold winters and factory life in the East. Pasadena held the promise of a healthier, more enjoyable way of life. They said, 'We'll show those easterners how to live.'"

"Amen," I say to Becky. Driving around Pasadena with the car windows wide open in February—I completely get this.

According to Becky, the railroads and the real estate industries trumpeted the attributes of warm weather, good health and relaxation to attract people to California. As a marketing ploy, they loaded trains with oranges and tossed them out to people across the country, saying, "See, this is what we're doing in California. Come join us." It was the beginning of what Becky calls a "new leisurism" that would eventually lead to the movie industry, luxurious beach houses and California dreaming. Pasadena symbolized a healthier, more hedonistic approach to life. Leave the Protestant work ethic in the East.

It was thanks to this new leisurism that the Arts and Crafts movement was able to thrive in California. I am in Bungalow Heaven here. Literally—that's the name of the neighbourhood we're now driving through, with blocks of one-storey homes with wide verandas, cedar shingle walls and muscular wood beams. Architecturally, the houses were designed to meld with the surrounding landscape, but Arts and Crafts was a way of living too, and aimed to promote artistic growth and physical well-being. It landed right in the sweet spot of a California that was seeking a new humanism.

Among the new settlers to Pasadena were David Berry Gamble, a second-generation member of the Procter & Gamble Company in Cincinnati, and his wife, Mary Huggins Gamble. In retirement they had spent a few winters here, and in 1907 decided to build a winter residence. They hired the firm of Greene and Greene to be their architects, and the result was the Gamble House.

It's no bungalow, but it is nearby.

Becky tells me to take a left and then a right. Before I know it, I am motoring along the affluent-looking Orange Grove Drive. I can barely keep my eyes on the road as my head swivels right and left taking in the lush landscape of citrus trees and noble mansions. Then, a gentle curve in the road and suddenly it's there.

The Gamble House is stunning and larger than life. I have seen countless images of it in books, magazines and lecture theatres, but nothing has prepared me for seeing this architectural masterpiece in real life.

I park the car, and Becky and I get out. I cannot take my eyes off the house. I am transfixed. Sitting on a slight rise of land, surrounded by trees and with a view of distant mountains, the three-storey Gamble House is a magnificent combination of open verandas, massive wood columns and redwood shake walls. The overhanging roofs, shaded porches and wood rafters seem to flow over the house like a series of gentle waterfalls. It has been called a symphony of wood, and I concur. I get the feeling that I am in front of one of the great works of art of the twentieth century.

I am astonished by the superb craftsmanship of the wood members, the textures of timber shingles and the copper downspouts. The powerful use of exposed beams, river-rock foundations and chimneys contribute to its beefy structure. It is in complete contrast to the tall, pointy-roofed buildings of the Victorian era with their delicate gingerbread detailing.

One reason I'm awestruck is the commanding presence of the house on the land. Seeing it for the first time, it looks wider and more solid than I had expected. I reason that the strong horizontal lines of the roofs and balconies contribute to the breadth of the front facade. Even though it is an expansive 8,100 square feet in area, it is harmoniously integrated into the natural landscape and appears to grow out of the ground.

The Gamble House has a Japanese influence. In 1893, the architect brothers Charles Sumner Greene and Henry Mather Greene visited the World's Columbian Exposition in Chicago, where they marvelled at the elegant Ho-o-den, or Phoenix Palace, a replica of a Japanese temple. It made a strong and lasting impression. The temple also fascinated their contemporary, Frank Lloyd Wright, and its inspiration is evident in his brick and stucco houses. The Greenes were drawn to the Japanese post-and-beam construction, wood joints, predominant roof and deep overhangs, and these would become the trademarks of their own work.

I have toured temples and houses in Japan and found that although writers refer to the Greenes' and Wright's buildings as having "Japanese" elements, it would be more precise to say "Buddhist," with that religion's ideal, manifest in its buildings, of fostering a love of natural materials in harmony with their surroundings. It is these religious philosophies, imports from Japan, that were the true basis for Wright's American houses and the Greenes' Gamble House in Pasadena.

Becky and I pick up our picnic supplies and carry them up the brick path to the entrance of the Gamble House. A handcrafted composition of redwood shakes, teak entry doors, leaded glass in meandering patterns and a copper hooded lantern, the Arts and Crafts entrance at the Gamble House might be one of the most appealing in all America.

The front door features a beautiful leaded art glass rendering of a gnarled oak tree, the "tree of life" designed by Charles Greene. The muted yellows and greens is a first taste of the themes of nature and craftsmanship that permeate the rest of the house.

I've always felt that a front door holds special significance as an architectural element. It is the point of welcome. It is the place of happy first hugs and greetings and, later, of long goodbyes. In my family we have countless photos of friends and family taken at a front door. It is the point of transition from the hostile outside world to warmth and protection, food and shelter.

We are met outside the door by docent Sheryl Scott, marketing manager of the Gamble House. She radiates positive energy. Sheryl tells us we will be cooking in the basement catering kitchen but, in the meantime, she invites us to look around the house.

Sheryl indicates the brass latch and invites me to swing open the broad entry door. We step from the sun-filled front terrace into a dark hallway where cool air gently strokes my face. The door closes behind us. Sheryl leaves us to attend to work but tells us she will join us later for lunch.

Once my eyes adapt to the near darkness, I can make out the main staircase that ascends from the entry hall to the second floor. Constructed of Burmese teak, it is a meticulous composition of interlocking rails, corbels, treads and risers inspired by Japanese wood joinery. I have seen photographs of this stairway in books, but seeing it in person is an immediate experience. There is no effort to hide the joints—rather, they are celebrated—and there are no sharp edges. Surfaces have a velvety finish and look like they've been sculpted out of butter. My breath is taken away by the level of craftsmanship in the fine detail. The Gamble House staircase is an Arts and Crafts masterpiece in itself.

I look behind me to admire subdued sunlight filtered through layers of the green and yellow art glass door panels. The soft glow of twisting branches, bark and leaves from the depicted tree brings the natural world into the house.

Becky and I examine the wood panels that line the walls of the vestibule. At first they all look like nothing more than wall panels, but I open one to reveal a clothes closet. A second, slightly larger panel turns out to be a door leading to the kitchen. It allowed servants to greet visitors at the front door and then literally disappear back into the woodwork.

"I feel like I'm walking through a work of art," says Becky as we move into the living room.

My eye is drawn to the broad fireplace that grounds the room under a massive carved wood truss. Its warm-toned wood mantel is surrounded by tiles inset with a pattern of flowers and vines. Glass-paned cabinet doors flank both sides.

All around the living room, above the picture rail, are California redwood panels carved in a Japanese style that showcases the natural grain of the wood. I point out that woven into the living room carpet is the tree of life, a repeating motif in the house. The room is finished with a collection of vases in a leafy-green glaze.

We float into the dining room, another space with great presence. It is anchored by a gleaming mahogany table, the edges of which are elegantly bevelled like the hand guard on a samurai sword. A chandelier of coloured glass and wood hangs on leather straps over the table and illuminates the room. Small light sconces hang on each side of a built-in sideboard for storing plates and utensils. Behind the sideboard, three stained glass panels in a pattern of winding red roses and vines emit an amber glow. Broad windows stretch along another wall to provide diners with a view of the garden. The carpet has a pattern of blue, beige and red flowers and vines.

I notice a discreet door at the rear and peek behind it. I find a small light-filled room that is the butler's pantry. We are entering the realm of the kitchen.

The Gamble House butler's pantry and kitchen are completely hidden away from sight. Families like the Gambles never went into the kitchen except to give instructions to the cook or housekeeper.

Becky and I step inside the butler's pantry. It is a pleasant space, filled with sunlight from the large double-hung sash windows. The windows are ajar, letting in fresh air to ventilate the eight-foot-square room. The two sets of maple cabinets that line the walls are precision-designed furniture with glass-faced doors that slide on wood tracks to save space. It's a well-laid-out room to maximize efficiency, so I judge it to be a comfortable working space for staff. Built into the sugar pine countertop are two nickel sinks with arched faucets. Stacked underneath the end counters are wide drawers containing tablecloths rolled to prevent creasing. I make a note of this excellent tip.

> *Even though the rest of the house is astoundingly beautiful in its Arts and Crafts glory, I am, as usual, happier poking around the kitchen.*

But to me, the most fascinating feature is the pass-through window from the kitchen. The pantry, situated as it is between the kitchen and the dining room, kept sounds and smells away from the dinner guests. When a dish was ready to be served, the cook would lift open the sliding wood panel, pop in the food and close it again. A lot of thought and precision had been put into sealing off the kitchen.

Finally we enter the kitchen of the Gamble House. I am taken aback by the generosity of space and bright sunlight. I tell Becky that I can see right away that the architects have designed the interconnected spaces for efficiency, easy movement and the comfort of the staff.

I immediately feel comfortable here. I can feel the muscles in my shoulders relax and already I am feeling hungry. Even though the rest of the house is astoundingly beautiful in its Arts and

Crafts glory, I am, as usual, happier poking around the kitchen. It is a utilitarian space for cooking, designed for work rather than to impress. I won't have to worry about leaving footprints on the antique Austrian rugs or knocking over and smashing the valuable ceramic vases. The only thing missing is a plate of chocolate chip cookies and a glass of milk.

Right away, I notice the lighter colour of the walls and wood details in the kitchen, the complete opposite look from the rest of the house. While dark teak and mahogany panelling dominate the public living areas, the kitchen cabinets are made of a light-coloured maple. It was a more practical and less expensive alternative. The difference is like night and day. In addition, the walls are clad in shiny white subway tiles for easy wipe-downs.

The space is anchored around a sturdy island table constructed from soft maple. There are three drawers, which open from either side and are lined with tin, to hold staples such as sugar, flour, salt and dry goods. This is the earliest kitchen island I have seen in a historic house. Multi-functional islands, as opposed to a plain table set in the middle of the room, did not catch on until decades later. This one, with the no-nonsense lines of an office desk, is visionary.

A gas stove from 1908 bears the proud nameplate of New Process. It is a six-burner beauty with temperature-control knobs, an oven and warming compartments. Above it is a ventilator with a broad steel-frame hood.

An "annunciator" box on the wall lists words—such as Living Room, Bedroom 1, Upstairs Hall, West Door—to indicate which room or door requires service. It makes me think of the bell system that summoned servants at Point Ellice House. The floor is covered in "battleship" linoleum in an attractive moss-green and beige check. Linoleum, invented in 1860, is a resilient, easy-to-clean and inexpensive floor covering ideal for such high-traffic areas as kitchens and hallways. Battleship linoleum, one of the heavier gauges, was originally manufactured to meet the specifications of the U.S. Navy for warship decks.

A door off the kitchen leads to the cold pantry, which houses the 1909 Eddy icebox and other storage cabinets for perishable food. Marble countertops in the pantry hold cooler temperatures, making them an ideal surface for rolling out dough in a warm climate.

My admiration for this kitchen increases even more when I find a "California cooler" mounted in the wall. This ingenious cabinet used the cool air flowing between the attic and the basement behind the wall to help keep fruit and vegetables fresh. As I open the cabinet door, I feel a gentle breeze on my hands and see wire racks stacked inside the wall. It is like preserving food inside a ventilation duct. I am amazed at the innovative methods to keep food cool before refrigeration.

The Gambles and Greenes extended the same high level of craftsmanship to the kitchen that they did to the public rooms, though with less exotic woods and in a simpler form. Upper cabinets are built with wood-frame joints that are held together with wood pegs, and the glass-faced doors have muntin patterns that echo the window elements throughout the rest of the house. Drawer handles on the lower cabinets have a rounded and smoothed finish so that they are a perfect fit for my fingers.

Becky and I experience the finishing touch on this cooking space when we walk into a neatly constructed wood porch at the rear of the kitchen. It is partially shaded by a canopy of orange trees. Large screened windows allow a breeze into the kitchen. Furnished with a table and chairs, it was an ideal place for staff and visiting workmen to gather and socialize.

"Most kitchens of the time were in the basement," I tell Becky. "But here, the staff had their own breezy terrace."

By the standards of the time, the Gambles went out of their way to create decent working conditions for their staff. And that probably benefited the Gambles too. Job opportunities such as office work were opening up to women whose choices might previously have been restricted to domestic service. A humane working kitchen was one way of hanging on to good help.

As our leisurely walkabout of the Gamble House ends at around ten thirty, Becky and I nod at each other, both of us thinking of our 1909 cooking task ahead. With our guests arriving at noon, we grab our cooler and lunch ingredients from where we left them at the front door and haul them downstairs to the catering kitchen in the basement.

A gentleman named Charles Perry, who is the president of the Culinary Historians of Southern California and a food critic, prepared me for this trip by sharing with me not only his thoughts on early twentieth-century Pasadena cuisine but also a copy of the 1905 *Los Angeles Times Cookbook—No. 2*, containing recipes sent in by readers. One of the chapters is dedicated solely to picnic fare.

In a nice touch, many of the recipes are credited with the names and addresses of the contributors. For today's picnic I settled on the following:

Ham and pickle sandwiches (Mrs. F. W. Koch, Etiwanda, Cal.)
Egg salad sandwiches (L.E.M., no address)
Chicken salad sandwiches (Miss Margaret W. Beckwith, Altadena, Cal.)
Automobile salad (Miss W. I. Puls, Riverside, Cal.)
Potato salad (Mrs. J. B. Kelsey, Palms, Cal.)
Fruit salad with oranges (Mrs. Frank Stone, Pasadena, Cal.)
California walnut pie with orange rind
Pink lemonade with berries
Malibu rosé wine

With a massive gas stove and a generous work table, the catering kitchen resembles the historic kitchen upstairs but is equipped with a modern refrigerator. I prepared the chicken and hard-boiled eggs the night before, so we get busy mashing eggs, slicing chicken thighs and chopping celery and onions to make the chicken salad and egg salad. I add my own touches of sliced grapes and walnuts to the chicken salad.

Becky and I combine slices of white and whole wheat bread to make contrasting colours. We cut away the crusts and spread on a firm layer of butter so the bread does not get soggy. Next, we spread the egg and chicken salads onto the bread slices. We have a good time experimenting with shapes and cuts before settling on small triangles. Becky artfully arranges our triangles close together and in different patterns inside wicker baskets with cloth napkins.

My mother was a skilled maker of festive party sandwiches, and I now try to emulate her pièce de résistance—the pinwheel. I take a stack of six slices of white bread, cut off the crusts and slather each slice with butter, mayonnaise and cream cheese. I make sure that the filling goes to the edges. I then place a slice of ham on each spread side and a gherkin pickle at the edge. Then I start to roll, all the while pressing down on the bread to help it glue together. And now for the fun part. I slice the roll into sections, creating perfect little pinwheels with green gherkin centres. It's a little like being a sushi chef.

As Becky garnishes the pinwheel sandwich basket with radishes and olives, she tells me that picnics were wildly popular in Pasadena in the early twentieth century. "When people from the East and the Midwest came to Los Angeles, they were pretty lonely. They kept in contact with each other and made new friends by forming state associations." By 1913, all forty-eight states were represented in Southern California.

The associations held meetings, shared information and, to help keep spirits up, organized annual picnics. The Iowa state association held the first Iowa Picnic in 1900, and it became an annual event in Pasadena. The largest picnic in early Pasadena, it boasted the presence of two to three thousand Iowans for years, a happy acknowledgement that, in California, you could eat outside during the winter months. (It's still going strong today, though in nearby Long Beach.) Another California tradition called Picnic Day started in 1909, at the University of California, Davis. It continues today and draws more than fifty thousand visitors.

Photographs of those earliest gatherings from the Pasadena Museum of History show people of all ages, the men in shirts and ties and boaters and the women in long dresses and graceful hats. The newspapers report that the organizers would make a roll call of cities, and residents would shout out when their city was announced.

There even exists an eye-popping photo showing hundreds of Winnipeggers picnicking in Brookside Park, Pasadena, in autumn 1923. Anyone familiar with the ferocity of a Winnipeg winter can sympathize with the happy smiles of those transplanted Canadians.

I wash my hands and jump into making salads. Becky and I are both intrigued by the one named "Automobile," which calls for tomatoes, lettuce, celery and pickled California olives in a mayonnaise-like boiled dressing. Southern California's first olive groves were planted by Spanish missionaries in 1769. The Mediterranean-like soil and climate proved ideal. Olive oil was first produced here commercially in the early 1800s. Over the years, the expansion of urban development caused olive farms to move to Northern California.

For dessert, I have found a fast and easy recipe for a traditional California walnut pie. I whisk together eggs, corn syrup, sugar, butter, grated orange peel, cinnamon, vanilla extract and chopped walnuts and pour the mixture into a prepared pie shell. It takes about five minutes to prepare and I pop it into the oven for forty minutes. I love simple. The kitchen fills with the smell of baking pie crust, toasting walnuts and burnt sugar.

Walnuts were first cultivated in California by Spanish Franciscan monks on their missions in the late 1700s. Unlike the varieties grown today, these "mission" walnuts were small, with hard shells. By the 1870s, English walnut trees were being grown in Southern California. And they grew and grew. Today, California walnuts account for 99 percent of the U.S. supply and three-quarters of world supply.

After a hectic ninety minutes, our picnic lunch is ready. I pull the walnut pie out of the oven and set it aside to cool.

Becky and I hustle our sandwich baskets and salad bowls up to the back terrace of the house. The garden, which is surrounded by a low clinker-brick wall, has a curving pond, brick paving and planters of vines and grasses. Paths paved with flat stones meander out across the lawns. It is a perfect place for a picnic.

Just as we are laying the table, Sheryl drifts onto the terrace with Angela George, the curator of Gamble House and Charles Perry. Mr. Perry and I have never met in person and I am very happy to shake the hand of this generous man.

We seat ourselves at the table, where Becky has done a beautiful job arranging the food. We have not done a bad job of making it, either.

The egg salad sandwich is perfect. Simple, wholesome—the filling combination of chopped eggs, mayonnaise, diced celery and red onion is nothing fancy. But I savour the rich and vibrant flavour. The textures in my mouth go all the way from creamy to crunchy. I am reminded that there is nothing more satisfying and wonderful than a flavourful egg salad sandwich on soft, fresh bread.

When I taste the chicken salad sandwich, I am taken back to my childhood and lunches in our family kitchen. My mom was a creative cook and she read as much as she could about food. Even while living in Toronto, she made chicken salad incorporating California grapes, walnuts, sweet relish and creamy mayo with the chicken chunks. It is great picnic food. I immediately reach for another. The grapes make it extra special.

Sheryl tries a pinwheel sandwich. "It's good," she says with surprise. "I would never have put ham and pickle together. But it's a great combination of sweet, smoky and salty."

As I expect everyone at the table has noticed by now, the sandwich fillings rely heavily on mayonnaise. Because of my low-fat fixation, I have not tasted mayonnaise for decades. The first mouthful of sandwich immediately brings back long-forgotten memories of my grandmother, who put mayo on every tomato and every vegetable she ate, every day.

The lunch conversation turns to why the Gambles would have wanted to build such a unique and innovative house. After all, the Gambles were conservative both politically and socially and were the kind of people who might have been expected to commission a building in Queen Anne or even neoclassical style. However, Mary Gamble had a passionate interest in art and design and she was not tied to convention.

Japanese-style gardens, architecture and artifacts were fashionable in Pasadena at the time because of trade across the Pacific. The foreign aesthetic caught the attention of the Gambles, and as an affluent and enlightened family, they were open to new experiences and tastes. They became infatuated with Japanese art and architecture, especially house design. Mary, who had been raised in Wisconsin by a wealthy and cultured family, worked closely with the Greene brothers on the design of the house as well as its furniture.

During the house's construction, the Gambles picked up and travelled to Japan. Few people visited Japan for pleasure at the time. It was more conventional for families to travel to England. But the Gambles went there and they loved it. They explored the country, its food and avant-garde arts and generally immersed themselves in its culture.

The automobile salad is passed around.

According to Charles, it is missing a vital ingredient—the one that explains the "automobile" in its name: a jar.

A jar allowed the salad to be easily transported in a Model T. At the time, only one in sixty Americans owned one of these first automobiles, so the term *jar salad* is a celebration of this new symbol of an eager, go-anywhere way of life.

I place a heaping portion on my plate and dive in. In her 1905 recipe, Miss W. I. Puls of Riverside, California, suggested that the salad could be garnished with red nasturtium blossoms—a flower that grows wild in California. I could not find any and so have substituted California pomegranate seeds to add a festive red brilliance. The lettuce, tomatoes and celery are crunchy and refreshing, and the pickled olives add a nice salty accent.

At last it is time for dessert. I bring out the toasty brown walnut pie, cut it into hunky slices and serve it with a berry and orange salad and a dollop of whipped cream.

We all bite into the pie at the same time, with a shared low moan. The heavenly melding of crunchy walnuts . . . an intense buttery sweet custard middle with hints of cinnamon and orange . . . the rich, flaky pie crust. Not to mention the berries and whipped cream.

"This pie is soooo good," says Becky, and no one disagrees with her.

Our picnic lunch is drawing to a close and I find myself in a blissful daze. There has been much chatter and laughter—and sufficient wine.

"We think that California dreaming is a recent thing," I find myself announcing to the guests. "But the Gambles were California dreaming at the turn of the century. Building a house with horizontal lines, unpainted wood and Japanese influences, enjoying glorious sunsets, sleeping on magnificent balconies at night."

Somehow, I can tell that nobody is listening. I look around the table.

Their eyes are saying: "That's nice, John. More pie, please."

Gamble House
4 Westmoreland Place, Pasadena, California
gamblehouse.org | (626) 793-3334

Dear Franny,

It was a thrill for me to tour the Gamble House and its kitchen. This was a big checkmark on my bucket list. I expected the magnificent woodwork. But for me, the big surprise was the kitchen and the commitment to innovative design and a progressive attitude toward the staff. Here are some Gamble House features that we could think about for our perfect kitchen.

The Greenes' central island is a real innovation for 1908. We might not have enough space to have an island, but for easier cooking, let's talk about having a drawer that opens from both sides.

Of course, light-coloured natural wood is a go-to choice for cupboards, but the Greenes took this to a new level. If we want to express warmth, simplicity, a connection with nature in our kitchen—the answer is wood.

Subway tiles are such a classic look—functional and conducive to the clean lines of the Arts and Crafts style. Fashions come and go, but subway tiles will be in style forever. The tiles at Gamble House rise up the wall behind the cabinets to allow for thorough cleaning.

Although there was art glass all over the Gamble House, there wasn't any in the kitchen. I would love to have a stained glass or art glass window in the kitchen to brighten up our days.

So nice to have a screened back porch off the kitchen! For coffee in the morning and casual dinners in the evening. Could life be any finer?

The Gamble House kitchen had a cold room where staff rolled out pastry for baking. The countertop was marble to keep the dough cool. Will this cause me to bake bread for you? Or a walnut pie?

Anything for you, my dear.

XOXO

J.

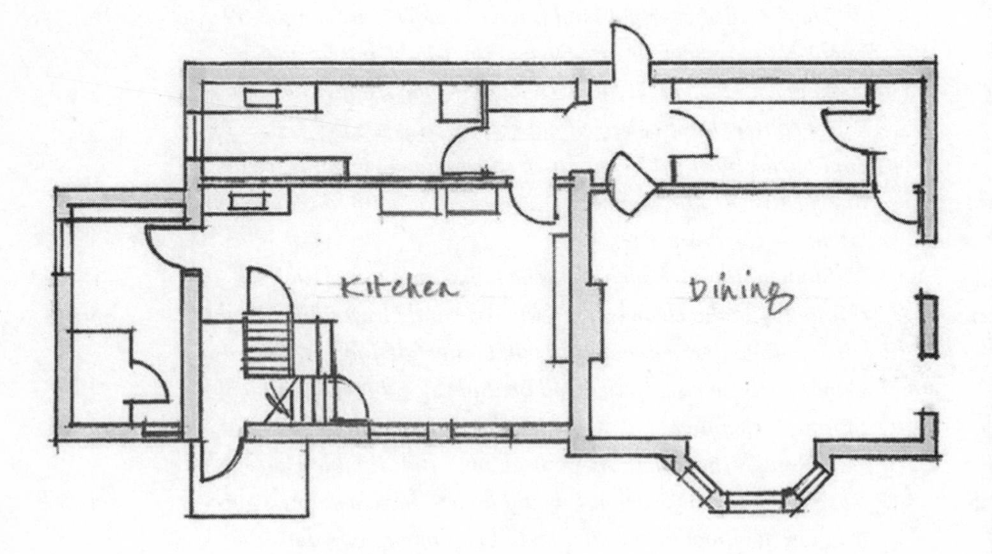

Plan Spadina Kitchen

interior Spadina Kitchen

7

SPADINA HOUSE KITCHEN, 1920

Toronto, Ontario

Even though I appreciate all architectural styles, I especially love the kitchens and interiors of Victorian houses. Spending time in the Victorian kitchen of Point Ellice House, I was captivated by the presence of the larder, scullery, coal stove and water filter, as well as the sheer numbers of pots, pans, irons, ladles and spoons that were hung on any available wall space or crammed on any available shelf. There was so much to take in. Yet I think the best time to see a Victorian house is Christmas. With so much attention paid to cooking, baking and entertaining at that time of year, the kitchen would be buzzing with activity. By introducing Christmas trees, Christmas crackers, Christmas cards and the tradition of rewarding children with gifts, the Victorians practically invented Christmas.

Spadina (pronounced Spa-*dee*-na) has been referred to as "the Victorian house par excellence" and as luck would have it, it is in Toronto, where I live. And I'm visiting at Christmas, my favourite holiday. Although I've visited the house before, I have never been at Spadina during the festive season. I have tried to be a good boy this year, so I am looking forward to having a Victorian yuletide experience.

With its laced-up corsets, starched collars and unspoken rules, Victorian society was the opposite of twenty-first-century high-tech, trash-talking North American pop culture. For the Victorians, tradition ruled, and they intertwined their social formality and their domestic architecture with textured wallpaper, heavy drapes and dainty hand-painted gold-leaf china teacups that I'm always terrified I will knock over and break.

More than all his finicky detail—not at all to my taste—I am intrigued by the relationship between the family and the servants; what it was like for both parties to be living under one roof, with the kitchen as staff-only territory.

Built in 1866, the fifty-five-room Spadina mansion, with its crystal chandeliers twinkling from the ceiling, oak balustrades sweeping up a grand stair and oriental carpets stretching across the drawing room, provided a splendid backdrop for the life of Albert and Mary Austin and their family. As president of the Consumers' Gas Company and the Dominion Bank, Albert was considered a professional member of the upper middle class. Mary was one of the founding members of the Toronto Symphony Orchestra and embraced a philanthropic role in the community. In 1984, the last members of the Austin family moved out of Spadina, and it was turned into a museum of posh family life in the early 1900s.

My mission today is to assist baker Ed Lyons in the Spadina kitchen. He will be preparing Scottish shortbread to be served at a Christmas celebration, and will be following a Victorian recipe. I have never made shortbread before and am eager to learn in the kitchen. As always, I arrive early to tour the mansion on my own before making my way to the kitchen. At the turn of the century I might have been greeted by servants at attention, dressed in black and white, bowing and curtseying to me as I came crunching up the snow-covered walk, but today I am on my own.

In the 1990s, I became aware of Spadina when I was a member of the Toronto Historical Board, with responsibility for setting corporate direction for the museum. Eventually, my term on the board expired and my visits became less frequent as other activities took over my life. But I always considered Spadina the premiere place to learn about Victorian life in Toronto. Even though the city has changed dramatically, the institutions, corporations and power structure of the city derive from the Victorian period and in some cases still exist today. I have not been back for a while, so this will be like visiting an old friend. (Yes, I tend to anthropomorphize historic houses.)

Surprisingly, at Spadina the visitors' entrance is in the humble basement, the realm of the servants, and so this is where I begin my tour. The drab, unadorned basement was used for washing linens, repairing clothes and storing tools and supplies. Spadina had a staff of around ten people at any one time, which would have included maids, a cook, a chauffeur, a gardener, a laundress/washer and a seamstress,

scullery

and this is where they spent much of their work time. I pass a room where preserves were once stored—jars of peaches, pears and currants and chili sauce made from tomatoes grown in the kitchen garden. The fact that this room was always kept under lock and key only underlines the strained and constricted relationships in the house. The family, it seems, did not trust the staff enough to not steal their jams.

Like Dorothy entering Oz, I ascend the narrow servants' stairs to the ground floor, and the house suddenly becomes a world teeming with oversized portraits, giant porcelain vases, elaborately carved wood furniture and walls covered in colourful flower-patterned wallpapers. As I step into the front hall, with its oriental carpets and elaborate crown mouldings, I see a curving grand staircase strung with holly and red ribbons. To my right and left are classical columns and doorways cloaked with heavy red velvet drapery leading to the drawing room and reception room. The hall is festooned with cedar boughs, wreaths, mistletoe and bright red poinsettias. In the distance, I can see a twelve-foot Christmas tree covered in sparkling ornaments and tinsel dominating the palm room.

The grandeur of the entrance hall leaves no doubt about the family's wealth and status. And clearly there were rules about what the visitor should see. The draperies act as picture frames that both direct my view and make sure I am impressed by what I am supposed to see, such as a marble fireplace in the drawing room and the vista through to the sunlit conservatory.

In the drawing room, with its gilt over-mantel mirrors, chaise seating and an 1888 Steinway piano, I see a room that was the centre of art, culture and conversation. In this room, Mary Austin hosted music recitals, threw parties and played the piano.

The reception room across the hall was used to host acquaintances who visited on "at home" day. As printed on their personal calling cards, the Austins were at home to receive visitors each Friday from 2 to 4 p.m. With visitors in mind, the room was swaddled in padded silk wallpaper and stuffed full of artifacts collected on trips around the world to act as conversation starters. The Austins always had pastries on hand to serve to guests at small, intimate parties in the reception room. At Christmas, visitors would sip mulled cider.

I make my way up the grand stair and find that the master bedroom is comfortable but modest in size and furnishings. While it was normal at the time for married couples of the upper middle class to sleep in separate rooms, the Austins had a double bed. The adjoining bathroom is modest too, save for a gas-powered shaving-cream warmer.

Wealthy couples kept separate bedrooms for a variety of reasons, one being that they had servants to help them dress, and it would not have been acceptable for a male servant to be in the bedroom while the lady of the house was being dressed. Sometimes these separate bedrooms had a private passage leading to a third bedroom where the couple could be together. They might then choose to return to their own rooms. It is formalities like these cumbersome sleeping arrangements that I find so fascinating about Victorian life.

If the legacy of Spadina is the transition from Victorian values and technology to a new modern age, then the most fascinating room in the house is the kitchen. I step carefully down the narrow servants' stairs to finally arrive at my destination.

Many late nineteenth-century kitchens were buried in the basement, but I am delighted to find that Spadina's—added to the house in 1898—is on the ground floor, at the rear of the house. Alongside the kitchen there is a scullery, the icebox room, a butler's pantry and a servery vestibule that leads into the dining room.

As soon as I step into the kitchen, it is instantly my favourite room in the house. Sparse, clean, functional, there is a basic honesty here—a perfect place for cooking.

Spadina's elaborate Victorian interiors are the yin to the austere kitchen's yang. While the rest of the house exemplifies nineteenth-century splendour and affluence, the kitchen has a no-nonsense look, with a linoleum floor of brown and blue squares and unadorned windows. In complete contrast to the other rooms, the walls are clad in simple wainscotting. High ceilings allow the hot air from cooking to rise, and the tall windows make for a surprisingly sunlit and airy space. The Spadina gardens are planted just outside, and on temperate days the open windows allow breezes to circulate and the scent of lilacs to waft in. There are no Christmas decorations in the kitchen.

Ed Lyons is waiting for me. He's an older, soft-spoken gentleman who has been baking, boiling and steaming in the Spadina kitchen for many years. As an expert in cooking techniques of the past, Ed is the ideal person to guide me in the ways of making Scottish shortbread. Even before I arrive, Ed has neatly arranged the bowls, cookie sheets, mixing spoons, flour sifter, rolling pin, spatula and cookie cutters on the large central wood work table.

"Hello. I'm John Ota. I'm here to help make cookies," I say, trying to make a little chit-chat.

"Nice to meet you," says Ed. "Let's get to work." I can tell that Ed is not a chit-chat guy.

Ed putters around, scooping flour out of a bin, measuring butter, filling a pitcher of water and fetching a bag of sugar. The kitchen is quiet except for the hissing of the historic gas oven preheating and Ed's feet shuffling on the floor.

While my mentor is engrossed in preparing for cookie making, I look about me and note how the house has been designed to keep the servants invisible—a recurring theme, like all the houses

I've seen so far that had servants. Whereas the front hall with its carved wood stairway is intended to make a dramatic impression on a visitor, the kitchen is tucked away at the back to hide service staff and deliveries. Staff toiled twelve to fourteen hours a day, cooking, cleaning, hauling buckets of hot water up and down stairs, answering the door, helping the family dress and always ready to attend to any need of the family or guests—but rarely be visible.

A wooden trellis screen covers the windows facing the street so that people outside could not see the cook at work. The Austins held elegant garden parties and an annual strawberry social and played croquet in the warmer months, and guests would be entertained on the lawns at the opposite side of the house. Servants cooked in the rear kitchen and then walked across the lawn to unobtrusively serve trays of strawberries and cream to guests, and then instantly disappeared from sight.

In addition, the servants were responsible for meeting the deliverymen at the back of the house and inspecting the meat and produce and making sure that everything had been delivered in the correct quantities. Pulling back a thin linen curtain and peeking out a back window, I can see in my mind's eye a delivery boy riding his bicycle up the rear walk with a basket on the handlebars loaded full of groceries.

The kitchen's back door opens directly onto the garden, where staff would have picked fresh vegetables for meals without being seen by the family or their guests. The Austins employed a gardener to grow vegetables and flowers and to tend to the greenhouse and three-acre grounds. They provided the gardener with a self-contained cottage on the grounds, where he lived with his family.

I read Ed's recipe for shortbread and begin to feel some apprehension. It comes from a volume called *Five Roses Cook Book,* published in 1915. I've never baked shortbread before and fear this antique recipe will produce something as bland as cardboard. The four ingredients are butter, sugar, flour and corn starch.

"Time to get to work!" announces Ed as he claps his hands twice. I feel that he has noticed that my mind is wandering. I realize that this is the second time he has directed this comment at me.

Ed fills a measuring cup with flour and slides the blunt edge of a knife across the top before he filters it through a hand-cranked sifter that removes any lumps from the flour and squeaks as he turns the handle. The flour drifts out the bottom of the sifter and Ed adds sugar to the bowl. As I watch him at work, I think about how he has baked these cookies using the same bowls, spoons and baking board countless times before and how he knows this historic kitchen like the back of his hand. He makes a point to show me that he is dredging candied fruit (his own innovation) in flour before he adds it to the bowl. Methodical, almost workmanlike, he moves competently to the next stage. As smoothly as a ballroom dancer, Ed fetches a bowl of one-inch butter cubes from the refrigerator.

"OK, John," says Ed. "Your job is to cut the butter into tiny pieces with two knives, so we can incorporate it with the dry ingredients to make the pastry."

Since I'd been anxiously waiting for something to do, I attack the butter with pent-up energy and excitement, a knife in each hand, slicing one blade in one direction and the other blade in the opposite direction. There is a cacophony of knives and bowl clanking together.

"Slow down, son, or you'll break it," advises Ed.

As I moderate my pace and cut the butter more carefully, my thoughts turn to the ovens where the shortbread will soon be baking. Two stoves sit side by side in the Spadina kitchen: the original cast-iron coal-burning stove and a later gas stove. Even though the old coal stove could be disassembled, it was never removed from the Spadina kitchen after the gas stove was installed in 1930. Like a stubborn employee who refuses to leave the office after they've been fired, it remains in the kitchen to this day.

Cooks had to be strong and energetic. They had to wear long sleeves to fend off the burns. I would be exhausted dealing with the coal stove countless times a day, not to mention many other physical cooking chores such as lifting heavy pots, hauling water and carrying platters of hot food.

I take a good long look at the intimidating heavy black coal stove. How did people cook on this thing? The stove is short, squat and covered in mysterious trap doors and compartments. Its little doors control the heat by adjusting air intake. I imagine that cooks were susceptible to burns, scarring and breathing in belching smoke. Not to mention setting the house on fire while lighting the coal. I'm glad that we're not using this small black contraption today because I don't want to cause an accident.

I almost miss seeing a finely detailed bronze match holder beside the coal stove that is the only hint of decoration on the bare walls of the kitchen.

Although the gas stove was invented in 1802 and was exhibited at the Great Exhibition of 1851 in London, it did not become widely used in Toronto until the 1900s, when large and reliable gas networks became available. The installation of the gas stove at Spadina in 1930 must have seemed a miraculous kitchen transformation.

Ed steps over to the gas stove and motions me to join him. "I'm going to preheat the oven to 350 degrees," he says. "This gas oven is a real beauty. Come and take a look at her."

I put down my bowl of sliced butter and move over to stand in front of the gas stove. It was built by the Moffat Company and is clad with white sheet metal and silvery chrome accents that allude to streamlined airplane design. My favourite touch is the small nameplate in italic script on the face of the stove that identifies it as the Miss Canada model. I can instantly see that the new gas stove was safer and more efficient than the coal-burning monster. It had six burners, two ovens and two warming ovens and featured temperature-control knobs rather than trap doors. If I were the Spadina cook dealing with that coal stove every day, I think I would have broken down in tears of gratitude at the arrival of Miss Canada.

The benefits of using the gas stove for the Spadina cook far outweighed wood and coal. The upkeep of a coal stove required about one hour a day to clean out ashes, carry coal and tend the fire. Add in the misery of dirt and smoke, the time to light the fire and break down coal, and the threat of being burned by a hot coal stove, and the arrival of the gas stove was a godsend.

For hundreds of years, somebody had to tend to the fire, prepare the food and manage the kitchen, and this all-day, every-day job fell to women. But now, with the invention of the gas stove, life for women changed dramatically. The ability to simply turn a knob on a metal box and immediately receive controllable heat for cooking was revolutionary. There was no longer a need to gather wood or coals, start a fire, stoke the fire and clean the ashes all in order to cook a meal.

So here's three cheers for the gas stove. Now most women had time to spend on matters other than just tending to the home fires.

Ed and I continue our dough making. He pours some water into the dry ingredients and then adds my sliced butter to the bowl. He demonstrates with his hands how to squeeze the ingredients together in the bowl to make a dough and then gives me a turn to get my hands into the act. After a minute or so Ed turns back to me and asks, "Do you think the dough's the right consistency to roll out?" Not having done this before, and being reluctant to admit it, I blurt out, "I guess so," which of course is of limited help.

Ed's next instruction is "OK, dust the board and rolling pin with flour." Then he tells me to take the ball of dough out of the bowl and roll it out on the floured board. The dough hangs together perfectly, and I roll it out gently as flat as a pancake. Ed next tells me, "Give it a little dusting of flour and then fold the pancake in half and roll it again. That'll get more air in the dough and make a lighter, crisper biscuit." I begin to feel a quiet sense of momentum and bonding with Ed. We're getting there. Soon, the dough, Ed and I will be ready for the cookie cutters.

Ed tells me that the linoleum floor has been replaced only once in the history of the kitchen and that the new floor pattern and material are based on the original. The battleship linoleum has a lively random pattern of beige, blue, maroon and yellow squares that complements the institutional mustard-coloured paint on the walls.

Ed hands me a few 1920s metal cookie cutters in the shapes of Christmas trees, stars and wreaths. They don't look any different from those made today. With the dough flat as a pancake and spread out on the board, Ed directs me to dip the cookie cutter in flour first to keep the dough from sticking to the metal. Of course, in my excitement and mental deafness I forget the flouring process precisely one second after his instruction. The dough stubbornly refuses to drop out of the cutter. I have to poke it out with my fingers, massacring the Christmas tree shape. It's a good thing I'm not doing brain surgery. Eventually I get the hang of it: dip the cutter in flour, press down firmly and quickly, don't wiggle the cutter or twist it, just pull it right out again, re-dip in flour . . . and we create a small army of cookies laid out on cookie sheets that Ed has lined with parchment paper.

Ed mentions that if we were making shortbread in the 1920s, the butter would have been kept cool in the oversized icebox that sits at the rear of the house in its own specially built room. Iceboxes were used from the mid-nineteenth century until the arrival of the refrigerator in the 1930s and stored perishables like milk, cream, butter, fish and vegetables from the garden. Many iceboxes were handsome pieces of furniture. Their walls were lined with layers of wool, felt, waterproof paper and air spaces.

We take a quick break and Ed takes me into the room that contains the icebox. The Spadina icebox is enormous. Fifteen feet high, four and a half feet wide and seven feet deep, it is four times the size of our refrigerator at home—hence the specially constructed room. Behind it, a meat locker was used that was large enough to hang hams and sides of beef and lamb to satisfy the meat-heavy diet of the Austins. Clearly, the Austins loved to eat and preferred not to worry about supplies running low.

According to culinary historian Fiona Lucas, the servants would load a tray at the top of the icebox with up to six hundred pounds of ice, and they were also responsible for ensuring that the melted runoff would be drained out to the garden for the flowers. The ice-on-top design allowed the heavier cold air to fall and circulate around the food compartments in the lower section.

In the quiet kitchen, the oven is hissing. "She's reached three-fifty," announces Ed. "Let's put the cookies in for thirty minutes and then we're done!"

Ed sets his hand timer and I give him the thumbs-up. I open the oven door and it squeaks and screeches in protest as if it hasn't been opened in decades. Ed slides the cookie sheets in and we look at each other. He gives me an impish wink.

While we wait for the cookies to bake, Ed shows me the scullery, adjacent to the kitchen. This was mainly used for washing dishes but also provided extra working space when the kitchen was busy. In houses built before indoor plumbing, the scullery used rainwater collected in barrels. The Spadina scullery contains a sink, storage shelves, work counters, tubs and buckets. The task of washing sets of dishes for the multiple family meals every day was solitary and labour intensive. At Spadina, the scullery maid was usually a young immigrant woman or a teenager from a farm in the country and she was at the bottom of the hierarchy of servants. After washing the dishes and silverware, she would carefully lay the dishes to dry on soft pads of towels to make sure they did not break. Ed explains that the cost of replacement dishes would have been deducted from her wages. He tells me that although her job wasn't easy, young women took it with the intention of being exposed to the manners of the upper classes and the possibility of moving up the house service ladder.

On the way back to the kitchen, we walk past the servery vestibule, the crucial passage to the formal dining room. It's equipped with a long plating table that allowed for last-minute garnishing before dishes were carried into the dining room. A small gas-operated 1908 hot plate could keep dishes warm before they were served. The swinging dining room door is padded with leather on the kitchen side to muffle sound and covered in a decorative fabric on the dining room side. The coverings are worn away at chest level on both sides of the door, the marks of where knuckles and elbows were used to open the door while balancing trays of food or dirty dishes. Scuff marks mar the metal trim at the bottom of the door, where heels and toes pushed the door open, ghosts of servants past.

As the cookies continue to bake, Ed and I take the dirty bowls to a sink at the rear of the kitchen that was reserved for washing pots and pans. Ed explains that in the 1920s, meat had more fat in it compared with today and that the drain of this sink was equipped with a grease catcher. A plumber would come by to extract the accumulated grease and sell it to grease collectors, who transformed it into products such as lipstick, soap, armaments and candles.

Kitchen cabinets—or built-in kitchen shelves on the walls with covering cupboard doors—were just coming into use in the 1910s and 1920s. The field of home economics enjoyed rising popularity in the early twentieth century, and with it an interest in kitchen efficiency and motion studies of housework. The first step towards built-ins came in the 1910s with the Hoosier cupboard, a free-standing unit designed with storage and counter space. We don't know when the cupboards at Spadina were installed, but the Austins were early adoptors of new technology, including radio and telephone.

A 1920s clock perched on a simple wood shelf between windows reminds us that the cookies should be done in ten minutes.

The kitchen cabinet doors at Spadina are faced with panes of glass that, unlike solid cupboard doors, lend a light airiness to the room and give the illusion of a larger kitchen. The glass allows for the viewing of rows of canned and packaged foods inside that is consistent with the belief that the kitchen was the most organized room in the house. Everything in the kitchen, it was thought, should appear neat, organized and tidy.

All the doors at Spadina are built with ventilating overhead windows, called transoms, above them, so the enticing aroma of our baking cookies swirls throughout the house. The transoms also allow light to enter the darker interior spaces. A 1920s clock perched on a simple wood shelf between windows reminds us that the cookies should be done in ten minutes. Ed is busily arranging the cooling racks on the centre work table.

Ed tells me that he and his wife, both retired, began volunteering in the Spadina kitchen after they replied to an ad in the newspaper. They are both food enthusiasts, so working in the kitchen had its appeal. I am staggered when he adds that he is eighty-eight years old. He hustles around the kitchen like someone forty years younger. All afternoon he has been zipping over to the oven, carrying the heavy flour bin to the table, moving cookie trays around and gathering up and washing bowls at the sink with gusto. I ask him his secret, and he tells me he goes to the gym at least five times a week.

Memo to self: Get to the gym more often.

The thirty minutes we wait for the cookies to bake feel like thirty hours. My stomach's growling. Ed checks the time, and finally it's time to check the cookies. The aroma of baking butter and sugar has driven me wild with cookie desire, and my mouth is watering. Ed grabs his pot holders, opens the squeaky oven door and pulls the cookie sheets out. They are golden brown and tempting to the eye, and Ed declares they are ready—not too pale, not too dark. He tips the pan and the cookies slide onto the cooling racks.

I feel a sense of satisfaction and amazement. The shortbread cookies, in the shapes of Christmas trees, stars and wreaths,

look divine. They smell heavenly. They taste even better. Warm, buttery and sweet, but not too sweet. The candied fruit gives them a light touch of extra sweetness and chewiness. I'm surprised, but I shouldn't be. Ingredients such as flour, butter and sugar are the fundamentals of baking. I learn (again) that just because a recipe is from another era, it doesn't mean the result won't be delicious for all time. These shortbreads are so good, in fact, that if I were stranded on a desert island, I'd be happy to eat them for the rest of my life.

We have baked three dozen buttery and crispy cookies. I must call upon every measure of willpower in my body to stop after eating three. After all, these cookies are meant for all of the visitors to Spadina, not just me. I stick my hands deep into my pockets as a form of self-inflicted cookie handcuffs. Even after I stop, the maddening aroma of freshly baked shortbread fills the kitchen and I yearn for more.

I shake hands with Ed and pat him on the back to celebrate our success. Ed then coolly removes his apron, washes his hands and gets ready to greet visitors to the museum and offer them our shortbread. As I look down at myself, I notice that my pants, sweater, hands and arms have turned white. I am completely covered in flour.

Spadina Museum: Historic House and Gardens
(also known as Spadina House)
285 Spadina Road, Toronto, Ontario
Call for hours and admission fees.
www.toronto.ca/spadina | spadina@toronto.ca | (416) 392-6910

Dear Franny,

Even though Spadina is a Victorian house, it offers several things for us to think about for our kitchen.

After a hundred-year absence, the butler's pantry is back. This would boost our creative cooking. A small separate room off the kitchen can act as a second kitchen, kitchen storage, work room, home office, powder room or even a cookbook library.

If we want to be a little different, for a low cost, linoleum can inject individuality, zest and colour into the kitchen. Visually attractive, easier on the feet and durable, it comes in varied and artistic patterns.

A contemporary reinterpretation of wainscoting could give proportion to the walls and protect them from scrapes and wear. A simple chair rail or unobtrusive, monochromatic wood facing.

Generous, tall windows and transoms will add good mood and humour to any kitchen. Transoms provide cross-ventilation and still allow for privacy. They also add a certain sense of romance, don't you think?

From the kitchen at Spadina you can slip outside to pick fresh vegetables and herbs from the vegetable garden. What could be nicer than a fresh sprig of parsley or a twig of rosemary picked from the herb garden to garnish the roast potatoes?

The passageway between the kitchen and the dining room has a table that can fold up to accommodate serving dishes and then fold down when it's not in use. This simple addition could almost be like having another set of hands.

As for servants, well, I don't see that happening. Unfortunately for you, I'm as good as you're going to get right now.

XOXO

J.

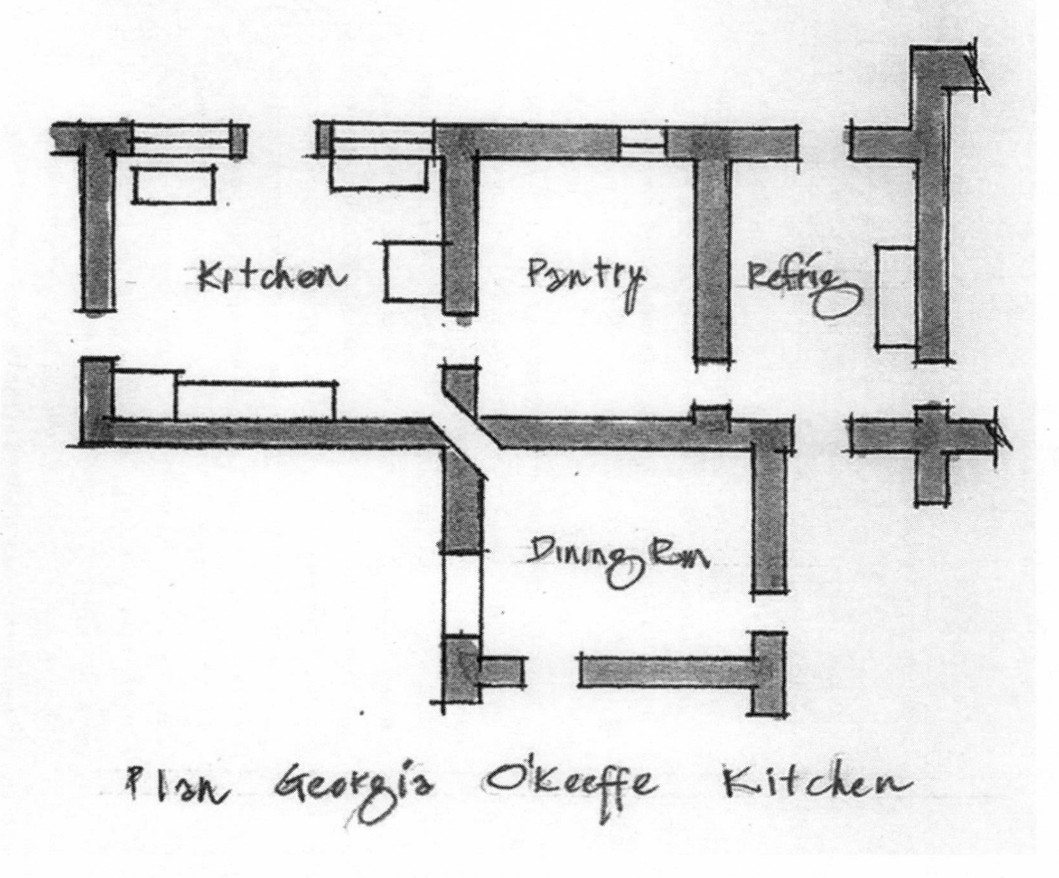
Kitchen
Pantry
Refrig
Dining Rm
Plan Georgia O'Keeffe Kitchen

Interior Georgia O'Keeffe Kitchen

8

GEORGIA O'KEEFFE KITCHEN, 1949

Abiquiu, New Mexico

Georgia O'Keeffe is one of my all-time favourite artists. I've always been drawn to her paintings of oversized irises, hollyhocks and lilies, but also of course those of buildings—houses, barns and New York skyscrapers.

I am most attracted to O'Keeffe's use of colour. I am a colour guy. Colour in clothes and interiors and nature—whatever—lifts my spirits. Seeing O'Keeffe's soft pinks, sky blues and bright yellows is one of my great pleasures in life. A thrill for me was to see her *Iris*, a 1929 oil on canvas, in person in Toronto. Franny and I have the poster in our house, but nothing is like seeing the real thing—the long, continuous brush strokes on the canvas and the vibrancy of the deep mauve petals and chartreuse leaves.

O'Keeffe's paintings trigger a positive emotional response in my body. The glow of her colours assures me that the world is good.

As a lark, I went online and searched for "Georgia O'Keeffe + kitchen." The result was a delightful surprise. O'Keeffe loved to cook and had a classic 1950s kitchen in her house in Abiquiu, New Mexico.

I did more digging and found a cookery school in Santa Fe that offered a class in cooking from Georgia O'Keeffe recipes.

That is why I am now driving north out of Albuquerque into Georgia O'Keeffe country, surrounded by hills of sand, sagebrush and a big New Mexico sky. I am thrilled to have my younger brother Chris with me on this adventure. Chris works in marketing for a major chocolate corporation based in Los Angeles. But more importantly to me, he is a foodie and my brother. Some brothers get together to golf, others to fish. We're together to explore Georgia O'Keeffe's recipes and kitchen.

Windows down all the way, we laugh about almost forgotten stories, catch up on our lives and reminisce about family dinners long past. As the car speeds past red cliffs, grand mesas and dry riverbeds, I am in awe of the magnificent southwestern landscape. I can almost hear the soulful voice of Ray Charles singing "America the Beautiful."

Our conversation turns to Georgia O'Keeffe and her move from downtown Manhattan to the kind of isolated landscape we are now driving through. "What was she doing out here?" wonders Chris. "What was she thinking?"

To help answer this question, we head first to the Georgia O'Keeffe Museum, in Santa Fe. The adobe-style building has a good selection of paintings from O'Keeffe's career, including close-ups of flower blossoms, waterfalls in the forest and charcoal drawings of New York. We learn that O'Keeffe was raised on a farm in Sun Prairie, Wisconsin, in 1887, where she learned about the natural world and gained an appreciation of the land. As a child she developed an interest in drawing and sketching and then moved to Chicago when she was seventeen to study at the School of the Art Institute. In 1918, she moved to New York at the encouragement of Alfred Stieglitz, who was captivated by her art. She married Stieglitz in 1924, and he organized annual shows in the city to exhibit her works. O'Keeffe had a highly successful career in New York and in the 1920s was recognized as one of America's most important artists.

But life was not all rosy. Even though she was deeply attached to Stieglitz, O'Keeffe found that her art and independence were weakened by being with him all the time. He led a life of continuous visits by friends and extended family while she craved quiet and seclusion to do her painting.

To escape Manhattan, O'Keeffe began visiting New Mexico in 1929 and she fell in love with the stark landscape of the Southwest. New Mexico gave her new colourful scenes to paint and the isolation to work without distraction. In 1949, three years after Stieglitz's death in New York, O'Keeffe moved permanently to the Southwest to begin a new life alone. She renovated an adobe house and lived a solitary life painting the New Mexico mountains.

As we walk around the gallery, I am most drawn to her paintings of New Mexico, especially *Untitled (Red and Yellow Cliffs)*, from 1940, depicting the landscape near her house in Abiquiu. I like the undulating red and pink forms that fade into each other to depict a mountain range. I am not an art expert, but even I can see a more abstract, simplified approach after her move to New Mexico. O'Keeffe was rejuvenated by the exotic

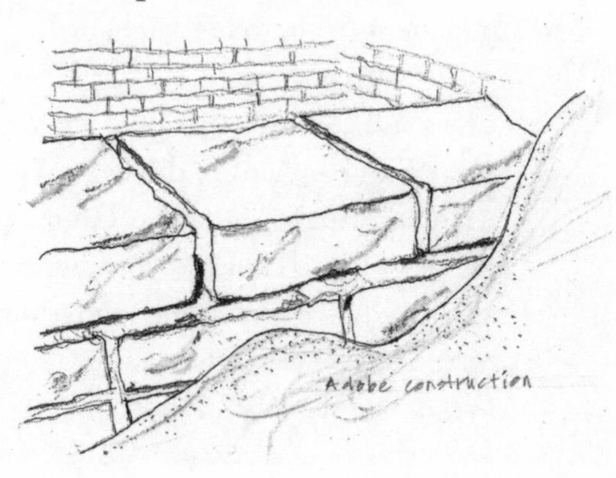

subject matter of endless badlands, enormous skies, sun-bleached animal skulls and bones, and a backdrop of ochres, oranges, reds and purples. These are her strongest works.

We come across a display of Georgia O'Keeffe's paintbox. The lid is open, and inside are bundles of her paintbrushes, squeezed tubes of paint, pastels and chalks, all neatly organized in compartments. The box is scuffed up and the handle is wrapped in masking tape. I picture O'Keeffe sitting on a ledge in the New Mexico mountains, painting a landscape with this paintbox by her side.

But the biggest surprise to me are photographs of her living at her house in Abiquiu. I am used to seeing O'Keeffe posing for a photographer, formal and with a stern face. These photos are different. There is a picture of her in jeans straddling a motorcycle, wearing goggles and a big smile. In another photo, she is casually hosting a meal on the patio outside her house, and in another she is lunching over beers and soup with friends at her dining room table. My favourite photograph captures her clad in an apron, hand on hip, cooking over her stove.

I point out the photographs to Chris. "She looks perfectly relaxed and happy."

Chris and I stare into the photographs and we agree. She was in her element here.

The photographs of O'Keeffe at her house only make me more impatient to get to her kitchen.

Chris and I jump back in the car and head north on Highway 84. After about an hour of driving, we pull into the small hamlet of Abiquiu, which is hidden off the main highway. The town of 250 people is a collection of trailers, small houses, a school and an old adobe church. The road is unpaved and dusty, and there's nothing else around for miles. Georgia O'Keeffe couldn't have found a more isolated spot.

We board a shuttle that takes us to O'Keeffe's house. We ride up a steep incline, turn into a gravel driveway and, suddenly, there it is.

"This looks like something out of a western movie," says Chris as we climb out of the shuttle.

O'Keeffe's house is a low, rambling, one-storey adobe perched at the top of a hill. Apricot trees, prickly pear cactus, sage and yucca grace the entrance. The expansive pink walls make me think of her minimalist paintings—plain canvases of soft colour.

I notice that O'Keeffe's dwelling features some distinctively southwestern elements: curvy shapes at the top of the parapets; a rusticated wood ladder leaning against a wall to provide access to the flat roof; elongated scuppers to drain water from the flat roof and shoot it out into the garden.

Built in 1740, the house looks as though it has grown out of the red earth, and in a way it has. Adobe bricks, which have been used in New Mexico for centuries, are made from clay and then laid to make a wall, with the same wet mud used as the mortar. Using the same clay for bricks and mortar brings a consistency to the buildings. Adobe bricks and mortar are not that strong, so the walls are built thick to carry the weight of a roof. In a multi-storey building, walls can be two feet thick. I am certain that O'Keeffe would have appreciated that the thick walls also cool the building in summer and absorb the sun's rays to warm it in winter.

After O'Keeffe bought the house in 1945, she renovated and added to it so that it became a nine-thousand-square-foot studio, courtyard, kitchen, living and eating space. Traditional adobe houses usually have only a few small windows, but O'Keeffe created a number of wide openings in the walls to allow for views to the hills. Nevertheless, she maintained the historical integrity of the house. It is still eighteenth-century adobe, but with touches of modernism.

At the rear of the building Chris and I find an enormous garden surrounded by a meandering pink wall. To me it looks like a small farm. I think to myself that even though this is a surprise, it makes sense—her roots were in agrarian Wisconsin. It is early spring, but already gardeners are tending to sprouting cabbages, tomatoes, onions, lettuce, green beans, corn, chives and chilies. One gardener who is digging up earth tells me that although the vegetable plants are most plentiful, they share space with herbs, flowers, trees, shrubs and raspberry bushes.

Chris points out some unusual stone-lined ditches that run the perimeter of the garden plots, and the gardener tells us they are irrigation ditches that bring water down from a mountaintop reservoir. It is a method of water transport that has been used in New Mexico for hundreds of years. Each Monday morning at ten, O'Keeffe's water rights would kick in, and the garden would be flooded like a rice paddy to create a desert oasis.

"O'Keeffe was so proud of her garden," she says. "She had to force herself to leave the house or she wouldn't get any painting done."

We step through a weathered wood door into a central courtyard, and I feel like we've entered the inner sanctum. I am surprised by the austerity of the courtyard—plain adobe walls, dirt floor and sage bushes. A collection of white elk skulls, antlers, driftwood and large rocks sits in a sheltered passageway off the courtyard. I recognize a patio door that O'Keeffe painted multiple times in different variations—with snowflakes or stepping stones.

I recognize a patio door that O'Keeffe painted multiple times in different variations—with snowflakes or stepping stones.

On display is a photo of O'Keeffe in the courtyard with two pet chow dogs she called her "little people." "I wonder if Georgia allowed them in the house," says Chris. We look each other in the eye and then crack up laughing. Growing up, we also had two dogs and they were allowed everywhere in the house. They slept every night in our parents' bed.

I get my first taste of an O'Keeffe interior when we enter her light-filled studio. Today it is a sitting room, with bare white walls and a stunning view out to the mountains. Chris and I study a photograph of the studio during O'Keeffe's painting heyday. The photo shows the room with two long work tables and a window shelf that held her paints and brushes. There are no paintings in the studio. O'Keeffe intentionally kept her rooms sparse and her walls white to encourage creativity and to help her envision her next painting.

We amble over to O'Keeffe's elegantly cool living room. The focal point is a gnarled tamarisk tree that grows in the garden and is framed in a Japanese-inspired screen window stretching across the back wall. The room is bathed in soft sunlight from four skylights that poke through the log-and-plank ceiling. O'Keeffe has furnished the room with a sharp-edged glass coffee table and striking black upholstered chairs that make me think of a Robert Motherwell painting. Splashes of colour come from pillows, potted plants and rugs. A spherical Noguchi lamp hangs from the ceiling, a gift from the sculptor during his visit to see O'Keeffe. The only painting in the room, *Above the Clouds*, gives the room a floating, expansive aura.

Our next stop is O'Keeffe's kitchen. As we step through the door, Chris and I turn to each other and exclaim a simultaneous "Wow!" We are immediately captivated by an astounding view through the wall-to-wall windows of the earth-red New Mexico mountains. It is almost as though we are standing outside on a rock outcropping and about to walk into the mesas and valleys. This dramatic vista was clearly an inspiration to O'Keeffe's art, and one she gazed at every day as she cooked and ate.

"Her kitchen is as minimal as some of her paintings," I say to Chris. A band of white-framed windows of varying heights and a glass door stretch along one wall of the twelve-by-sixteen-foot room. The walls are painted white, with two adobe-coloured (taupe) accent walls that echo rectangular forms in her art. The floor is covered in a taupe linoleum that provides a consistent colour palette. The ceiling of natural wood logs and planking adds a rustic ambience and reflects local New Mexico building methods. The kitchen is bathed in bright sunlight from the windows. I adore the subtle play of light and shadow on the curving surfaces of the adobe walls.

Although the kitchen is early 1950s, the clean, modernist aesthetic is consistent with interior design today. Yet O'Keeffe's kitchen has a warmth that pure modernist interiors lack. She has transplanted a New York modernist kitchen into the natural New Mexico landscape that she loves and melded the best of both worlds. This kitchen feels healthy, creative and in tune with nature. It has a feeling of cleanliness, hope, possibilities and well-being. There is no art on the walls. But maybe that's no surprise. O'Keeffe let the kitchen itself be the art.

The eating table is a simple flat sheet of plywood supported by sawhorse legs. Designed by O'Keeffe, who liked to use plywood for tables, it is painted with a thin coat of opaque white paint, a pleasing whitewashed finish consistent with the clean look of the room. O'Keeffe plywood tables are also in the book room, dining room and studio.

The white plywood makes me recall the whitewash that early settlers used to paint the walls of their houses and barns. I find myself attracted to whitewash, not only for its historical associations but also because it gives a soft, blanched look to wood and still allows the grain to come through. Whitewash adds serenity to this kitchen and enhances the soft white-on-white tones of the interior.

Simple wood chairs and a bench, also gently whitewashed, are positioned in front of the kitchen windows. I can almost see O'Keeffe sitting here and gazing out the windows while she enjoyed her meals. Her 1950s Philco radio sits on the table and would have provided music as she cooked her meals. I want to switch it on, but I hold myself back. Her pots of large geraniums sit on a wide window ledge.

I smile as my eyes land on a couch against the kitchen wall, covered with a white sheet. I can see that O'Keeffe entertained in her kitchen and provided her guests with a comfortable place to sit as she made dinner. None of this "isolated woman stuck in the kitchen" for Georgia O'Keeffe.

The kitchen is equipped with a Chamber gas stove, a Kenmore dishwasher and a set of drawers and countertop—all in white. Even the dustpan hanging at the end of the counter is white. There are no upper cabinets on the wall, and that gives the room a more open feeling. By the 1940's, built-in kitchen cabinets had become a "must-have" in most North American kitchen—but not for Georgia O'Keeffe, who had a very particular eye. The sink area is tidy, with only a dish rack, fire extinguisher and paper towel holder on the counter. It looks as though O'Keeffe just stepped out of the kitchen to pull some carrots from the garden.

I show Chris two sets of white metal cupboards set into the wall beside the stove. O'Keeffe found them in the Sears catalogue; they are a perfect visual addition to the kitchen. The cupboards slip so seamlessly into the thick adobe wall that they look custom made for the house. In addition, the top edge of the cupboard doors has a soft curve that echoes the wavy adobe. They still hold the elegant martini glasses she used at her dinner parties.

An unusual appliance sits next to the couch: a white rectangular box on wheels. It is called a Kenmore mangle, and when I discern from the guide materials that O'Keeffe used it to iron her bedsheets and dresses and shirts, I am not surprised. With her propensity for painting broad, perfectly flat walls and planes of colour, clearly O'Keeffe did not tolerate wrinkled clothing, sheets, pillow covers and linens.

For me, the most poignant aspect of the kitchen is the bare light bulbs that hang on their wires, one over the table and one near the sink. These reflect O'Keeffe's dislike for unnecessary clutter and her preference for functional design. In fact, all the light bulbs in the house are uncovered except for the Noguchi lamp in the living room. Clearly no lampshades—other than a pure Noguchi lamp—met O'Keeffe's refined standards. She would rather go bare.

Behind a door next to the stove is a ten-by-twelve-foot pantry. In my journey to find the perfect kitchen I have seen several pantries, and to my mind Georgia O'Keeffe's wins the prize for Best Pantry in North America. It is not just the size of the room or that it's so well stocked. The pantry reverberates with an intangible passion for cooking, gardening and eating. It is a reflection of the big part that food played in O'Keeffe's life.

Like her art, Georgia O'Keeffe's pantry is orderly and precise. The open shelves tell a story about her eating preferences and habits. All four walls contain her pots, pans, glassware, spice jars, coffee (Taster's Choice), coffeepots and coffee grinders neatly organized on white-painted wood shelves. Baskets for picking raspberries hang from the ceiling, and a large section of one wall is dedicated to drying herbs. Multiple tea pots and containers of tea line the shelves. There are sections for roasting pans, casserole dishes by Corning, pots by Le Creuset and a shelf for tins of McCormick spices. O'Keeffe kept her liquor and cooking wines in a locked cabinet and used a shelf-mounted ice crusher that resembles an old-fashioned hand-crank pencil sharpener.

"Look at this!" I say to Chris and point to O'Keeffe's electric frying pan on a pantry shelf. "It's the same as Mom's." We laugh with affection at the familiar sight of the pan that appeared day after day on our kitchen table sizzling up bacon and eggs, sukiyaki, Spanish rice and the occasional steak Diane. My mom would have loved to know that she and Georgia O'Keeffe cooked with the identical model.

There are few canned goods on the shelves. O'Keeffe grew her own fruits and vegetables in her organic garden and spent time preserving and freezing them, as evidenced by the cardboard boxes of jars and other containers in the pantry. I take note of a yogurt incubator on a shelf, an unusual appliance for the 1950s.

I realize that O'Keeffe's pantry is like her paintbox. Instead of paints and brushes organized into little compartments, her pantry shelves offer a palette of spices, herbs, flavours, cooking ingredients and cookware.

I poke my head through a doorway off the pantry and discover the refrigerator and freezer in the adjoining back porch. At a time when much of the rest of America was stocking up with convenience foods, O'Keeffe's garden provided all the fruits, herbs and vegetables she needed. The back porch was used for preserving, canning and freezing fruits and vegetables for winter months.

As we leave the kitchen, I spy the elegant beer steins that O'Keeffe was drinking from in one of the photographs at the museum. They are narrow at the base and graciously curve into a larger flute at the top. I can only guess that O'Keeffe, the consummate artist and designer, chose the fluted shape to blend with the gentle curves in her adobe walls.

Chris and I attend a Georgia O'Keeffe–inspired cooking class at the Santa Fe School of Cooking, which is just steps away from the centre of the historic downtown. I am so excited about attending that I insist we arrive thirty minutes early. We spend time in the bookstore/market attached to the cooking school, browsing the ceramics, cookbooks, utensils, jellies and spices that all have a distinct Santa Fe herb fragrance to them. It's a good way to get warmed up for a cooking class.

Finally, the doors to the classroom open and, as Chris and I walk in along with about twenty other O'Keeffe devotees, my head swivels as I admire the shiny cooking studio. The front of the room is dominated by a long counter with a built-in range, multiple ovens, high-end appliances and overhead video screens. Classroom seating is provided at about fifteen dining tables that seat three or four students each. Sunlight fills the room from the floor-to-ceiling windows at the rear; the walls are southwestern shades of salmon and turquoise; gleaming cutlery sits on granite counters and fluted bottles of vinegars and oils are set out on the tabletops. It all creates an uplifting atmosphere for the O'Keeffe culinary demonstration. I am ready to cook.

The first speaker is Margaret Wood, the author of the cookbook *A Painter's Kitchen: Recipes from the Kitchen of Georgia O'Keeffe*. O'Keeffe was ninety when Wood, twenty-four, moved to New Mexico in 1977 to work as a cook and caretaker for the artist. O'Keeffe's eyesight was poor, limited mostly to peripheral vision, but she was in full command of her tastes. Wood remembers O'Keeffe as an artist but also as a beautiful older woman with an endearing laugh who was a knowledgeable gardener and a compassionate employer.

"Miss O'Keeffe was a lover of nutritious, tasty and simple food," Wood tells us. "She was picky about how her food was presented to her and she was also a health nut."

According to Wood, O'Keefe grew up on a farm eating fresh food and watching her family and others can, dry, and otherwise preserve food. She created her organic garden in Abiquiu in the late 1940s, using information gleaned from books and magazines written by Jerome Rodale, considered the "father of organic gardening" in the United States. (The Rodale family has since published numerous books about vegetables, herbs, and healthy lifestyles and in 1950 launched *Prevention* magazine, which continues today.)

O'Keeffe was an acquaintance and follower of pioneering nutritionist Adelle Davis, author of the 1954 book *Let's Eat Right to Keep Fit*. Davis advocated eating natural foods that were free of additives, a policy that O'Keeffe carefully followed. "Her garden was an extension of her kitchen," says Wood. "We ate raspberries, apples, pears, herbs and vegetables from the garden."

Wood prepared O'Keeffe's supper and breakfast, working a night shift that began at five in the afternoon and ended at nine in the morning. "Miss O'Keeffe taught me how to cook," she says. "She showed me how to properly pick and care for herbs and other greens. Whether it was lovage, basil, dill or mint, the herbs were washed, patted dry with paper towels, stored in the fridge in damp paper towels and eaten the same day."

Wood presents a bunch of green leaves to the class and tells us that it is lovage, O'Keeffe's favourite herb. She rubs the lovage between her fingers and then passes it around the class for us to smell.

"Miss O'Keeffe also liked wild vegetables, including fried flowers and dandelion greens mixed into mashed potatoes," says Wood. "She regularly asked me to pick wild watercress by the stream beds."

Foraging is now a fashionable culinary practice. But of course it was once a common, everyday activity (like baking your own bread or making your own yogurt or growing your own vegetables) before it briefly vanished after the war and then was revived as a trend. It was what humans did, and what most North Americans did until the postwar introduction of processed foods. So in that sense O'Keeffe—who was foraging in the 1950s—wasn't so much ahead of the times as she was old-fashioned, a holdout continuing a traditional way of eating—local, sustainable, seasonal—that had been a way of life for a long time.

The class continues as chef-instructor Allen Smith announces that the first dish to be made this morning is a Georgia O'Keeffe summer salad. Smith is an engaging chef with many years of experience teaching cooking and working as a chef at restaurants across the country.

"O'Keeffe's recipes are simple," he tells the class. "But they're delicious because they're made with fresh and high-quality ingredients." This makes me think about her art and her house. O'Keeffe's life is consistent.

I am so pumped that I am the first at our table to volunteer to chop the lettuce, arugula and lovage leaves that will make up the salad. Even though there are better-qualified cooks at our table, I can't help myself. I pick up a knife and cutting board and away I go. I am sure Chef Smith thinks I am a little deranged, but he doesn't say anything.

Over the next three hours, we chop and slice at our tables as Chef Smith talks about his views on the quality of eggs, beef, chicken, milk, olive oil and salt, all the while cooking Georgia O'Keeffe recipes before our eyes. He makes it clear to us that it is the quality and freshness of ingredients that is crucial to creating great dishes. He goes on a rant about his campaign against industrial farm–produced eggs and how he will only accept eggs from free-range chickens, for their deep yellow yolks. Chef Smith is emphatic that he will go to any length and expense to attain the freshest meat and vegetables. He uses only the best olive oils that he has sampled and specific seasonings from around the world to best accent his dishes.

I listen intently to the narration on high culinary standards. Periodically, Margaret Wood jumps into the conversation to share memories and recipes from her life with O'Keeffe. It is a lively show of cooking and storytelling. I am having the time of my life. I peek over at my brother, and he is engrossed in the cooking class as well.

Chef Smith instructs us on how to prepare O'Keeffe's salad dressing using olive oil, an equal amount of high-quality vegetable oil and some fresh lemon juice. We add Maille Dijon mustard and chopped garlic. Smith is particular about our using a lighter-flavoured olive oil from the south of Spain and the Maille Dijon mustard. Freshly chopped tarragon, dill, basil and flat-leaf parsley are added to the mixture. "Keep the dressing simple," advises Chef Smith. "Let the ingredients come through."

I have to fight Chris for the honour of tossing the salad at our table. Knowing it is Miss O'Keeffe's favourite herb, I add a generous amount of chopped lovage.

For the next dish, which is a corn soup, Smith instructs us to slice corn kernels off fresh cobs. Chris and I drop the kernels into a blender and add organic milk, minced onion and a special O'Keeffe soup mix of kelp, soy flour, brewer's yeast and powdered milk that was influenced by Adelle Davis. I am taken aback by O'Keeffe's healthy mix, and I cannot say I am looking forward to tasting a corn soup with brewer's yeast in it. But this is a Georgia O'Keeffe recipe. It will taste authentic, at least.

Margaret Wood tells us a story about how, when she arrived each day at the house, she would ask O'Keeffe what dish she'd like for dinner. The reply was often "Let's see what's ripe in the garden." Vegetables were usually steamed. O'Keeffe made sure that that much of the garden produce was dried, frozen or canned for winter use. Wood adds that O'Keeffe was not much of a drinker. As she says this, my mind goes back to the locked liquor cabinet.

For the main dish, Chef Smith has us prepare baked chicken with lemon. We form the chicken into what is called an airline cut or suprême, where we flatten the breast slightly and leave the wing bone in. Smith explains that this is a cut that airlines used to serve in the 1950s and '60s, when air travel was an elite form of transportation, stewardesses wore crisp, tailored suits and food was served on white china with silverware. We coat the chicken with salt, pepper, lemon juice and garlic and slide a thin slice of lemon under the skin. The chicken breasts are baked in a casserole dish, to be served with a side of fried onions, potatoes and bundles of asparagus tied in green string.

As much as O'Keeffe loved vegetables, she was not a vegetarian. She ate chicken, enchiladas with ground beef and a thick, juicy steak every week or two. She preferred organic meats and grains, and bought from local farmers. She ate homemade bread, her eggs and honey were bought from neighbours, and she made—as I saw—her own yogurt.

I am very happy when Chef Smith announces that for dessert we'll be making Norwegian apple pie cake with rum sauce, an O'Keeffe favourite. Anything with fruit, cake and rum sauce is great news to me. In fact, I don't even need to have the fruit.

Chef Smith shows us how to make the cake with sugar, butter and freshly picked Granny Smith apples. Chris and I pare the shiny bright green fruit at a furious rate to keep up with the rest of the class.

"I love Granny Smith apples," I say to Chris as we slice our apples into thin wedges. "I've got a good feeling about this cake."

Smith tells us that the key to making great apple desserts is to use fresh-picked apples. I am stunned when he tells us that most apples are picked in the fall and then kept in cold storage all year, so that often we are biting into something that's months old. He tells us that the apples in our hands are favourites of his, from New York State, and have only just been harvested.

Margaret Wood tells us that the aesthetics of a meal were important to O'Keeffe. Her whitewashed plywood dining table at Abiquiu was set with plain white china, straw mats in natural or bright colours, fringed white cotton napkins and stainless steel silverware. "It was not quite minimalist, but it was very simple," Wood says. Similarly, with her food, "the focus was not on what kind of olive oil or what kind of pepper she used. The focus was on the quality of the food."

I am famished. By the time the cake comes out of the oven it is after noon. I am ready to eat. Sitting, watching, inhaling the aromas of the cooking class, I am practically eating my pencil. As I place food on my plate, I try hard not to appear like a glutton.

First into my mouth is the summer salad. Fresh and flavourful, the herbs come through, especially O'Keeffe's favourite lovage, tasting like celery. The greens are enhanced by the special olive oil, mustard and coarse sea salt. Chris and I agree that this is the best-tasting salad we've ever had.

A real shocker is the corn soup. I am infatuated with it. I sip it slowly to savour the intense sweet corn essence and enjoy it as long as I can. My concerns about O'Keeffe's healthy additions are unfounded. The soup is thinner than I expected but has more flavour than I could ever have imagined. It is corn nectar. A touch of paprika and chives adds a final New Mexico flourish.

The airline-cut roast chicken looks, smells and tastes terrific. The lemon slice under the skin adds zing to the tender and juicy meat.

I really like to have the attached wing to gnaw on. It's a chicken any Pan Am stewardess of the '60s would have been proud to serve.

The apple pie cake is the perfect ending to a superb O'Keeffe dinner. It's pleasant tartness is countered by the sweetness of the butter rum sauce, and I am in heaven. Just as I had hoped, the apple has a bright and tangy flavour that could only come from freshly picked fruit. But it gets even better when we add a large dollop of whipped cream. As I spoon the exquisite dessert into my mouth, I experience a vision of the apple trees in O'Keeffe's garden. Butter, sugar, apples and whipped cream—I can't believe that the famous artist could cook something like this.

Chris and I scrape the bottom of our bowls to extract any last trace of cake, sauce and cream. We agree that it has been an unforgettable trip to New Mexico. Besides creating a whole new body of art work here, O'Keeffe gloried in her garden and a life of traditional cooking and eating that stemmed from her childhood. I think you get to know a person much better when you sit in their kitchen and taste what they're eating and see how they live their life. It looks to me like her years at Abiquiu might have been the happiest time of her life.

Before we left Abiquiu, the tour guides recommended that we see the nearby New Mexico mountain ranges in late afternoon. They said that's when the setting sun turns the mountains into strikingly beautiful shades of scarlet, orange and pink.

We went and the mountains were ablaze.

They were an epiphany.

Georgia O'Keeffe Home and Studio

Tours begin at the Welcome Center at 21120 U.S. 84, Abiquiu, New Mexico. Reservations are essential.

www.okeeffemuseum.org | abiquiutours@gokm.org | (505) 946-1000

Dear Franny,

The big statement in O'Keeffe's kitchen is her love of nature. The kitchen focuses all attention on the panoramic view of the mountains. Her garden and the wild landscape were an extension of her kitchen. A love of nature should be the starting point for decisions about food, aesthetics and lifestyle in our kitchen.

I really liked the whitewashed plywood table in O'Keeffe's kitchen. Whitewash is a good combination of a modern white look with a historical wood-grain finish. It gives everything a clean appearance, which is just what the early settlers were after. We can whitewash the cupboards, the table and chairs—all the woodwork. Am I crazy? Don't answer that.

This is the first kitchen I've visited that has plants in it. O'Keeffe's big geraniums add splashes of green to the white-and-taupe colour scheme. Just imagine the vibrant reds when the geraniums bloom. Let's have some geraniums in the kitchen in tribute to Miss O'Keeffe.

O'Keeffe's über-pantry is big, but it's also as organized as her paintbox. Floor-to-ceiling shelves on all four walls and foraging baskets hanging from the ceiling. There's something wistful about walking a river and foraging for greens.

Plywood tabletop on sawhorses, no-nonsense sink and stove. O'Keeffe cooked healthy, tasty, creative food in a spectacular natural setting. She had no need for fancy-schmancy. I'm into that.

Oh, and by the way, my love—please, no lock on the liquor cabinet for me.

XOXO

J.

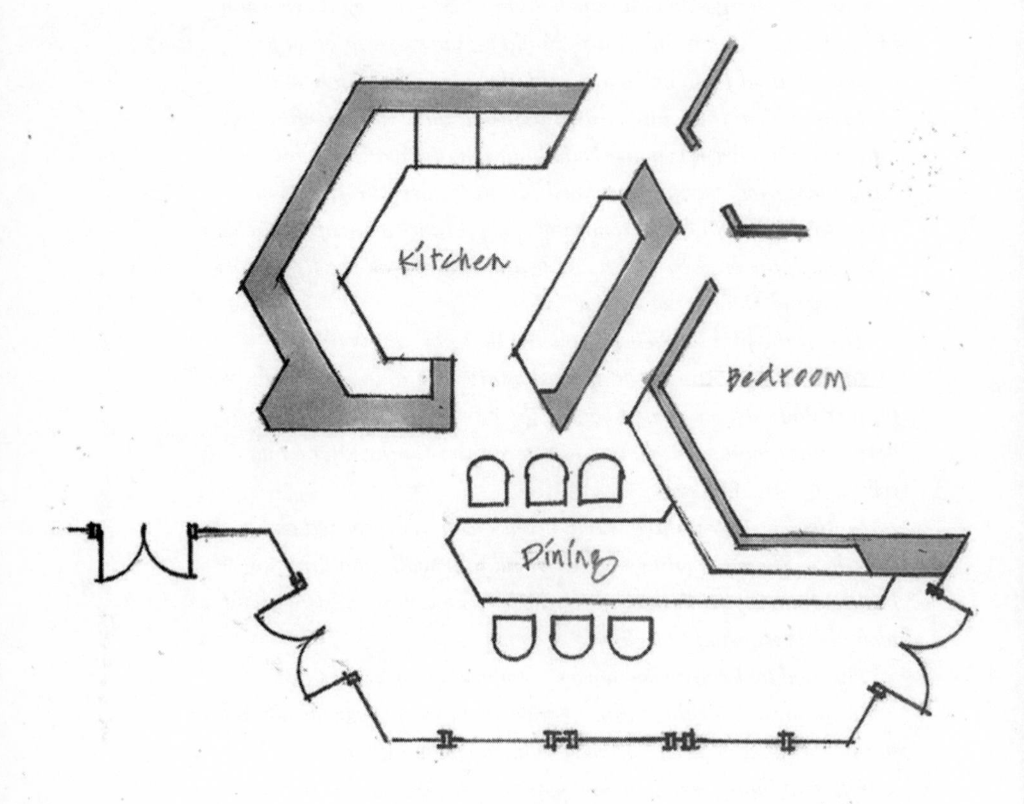

Plan Kentuck Knob Kitchen

Interior Kentuck Knob Kitchen

9

FRANK LLOYD WRIGHT KITCHEN at Kentuck Knob, 1956

Mill Run, Pennsylvania

During a visit to Boston, Massachusetts, I was fortunate to visit the John F. Kennedy Presidential Library and Museum and view its collection of photographs, film clips and documents of Kennedy's life and term in office from 1960 to 1963. I became intoxicated with Kennedy's altruism, energy and desire to change conventions that had existed for hundreds of years. He seemed to give America a new start.

What struck me most were his words. Smart, inspiring, visionary, his words convinced others that he could lead them to a better world.

My favourite Kennedy speech was the "We choose to go to the moon" one. Going to the moon was an outrageous thought in

1962. But Kennedy had a vision and he brought the rest of the world on-side. "Many years ago," he said in that famous speech, "the great British explorer George Mallory, who was to die on Mount Everest, was asked why did he want to climb it. He said, 'Because it is there.' Well, space is there, and we're going to climb it, and the moon and the planets are there, and new hopes for knowledge and peace are there. And, therefore, as we set sail we ask God's blessing on the most hazardous and dangerous and greatest adventure on which man has ever embarked."

I think Kennedy's words epitomize the spirit of the mid-century. The 1950s and '60s were a time of hope and prosperity, with a sense that life was going to continually get better. The sky was the limit.

Whenever I see mid-century modern architecture, interiors or furniture, I think of Kennedy's positive spirit. It is the architecture of optimism.

During the late 1950s and early '60s, America was obsessed with outer space. Even before Kennedy's speech, America and the USSR were in a space race. In the late '50s, it looked like the Soviets were winning. They launched a satellite named *Sputnik* and then Yuri Gagarin was the first human in space. On February 20, 1962, America gained the lead when John Glenn orbited the Earth three times. (My brother Chris was born that day. Mom said she and John Glenn were in orbit together.) The two nations were racing neck and neck.

The preoccupation with outer space was everywhere and it spilled over into architecture and design. The future was now and it was all

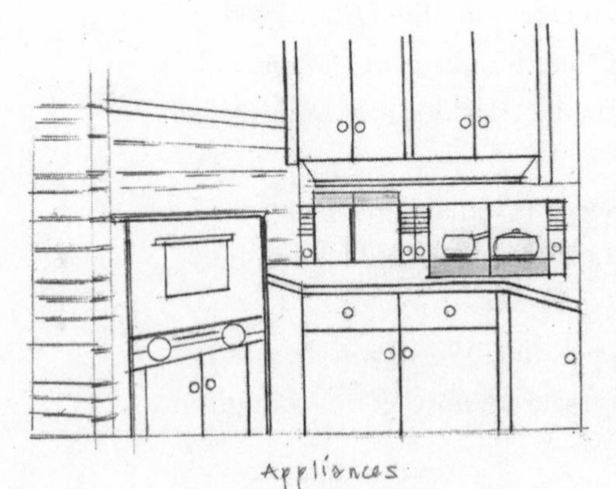
Appliances

about jets, rockets and aerospace design. Everything was made to look light and streamlined, as though it was about to rocket off into the universe. Cars, houses, coffee shops, cocktail lounges, motels and doughnut shops were designed in the futuristic style. The look was shiny chrome, blinking lights, spinning neon signs with starbursts and sleek rocket-ship fins.

Architects such as Richard Neutra, Pierre Koenig and John Lautner jumped at the chance to reinvent shapes and styles in their buildings. They broke all the rules. The new look was aerodynamic, including flying-saucer domes and angular roofs hovering above glass walls. It was the look of tomorrow.

How to Spot Mid-Century Modern:

- Angular roofs, open spaces, minimal ornamentation.
- Flat Planes, walls of glass.
- Step-down living rooms, split-level floors.
- Patios, courtyards: seamless indoor/outdoor life.
- Furniture of curving, clean lines.

Ultimately, mid-century modern was a style that designers hoped could lead to a better world. It reflected hopes and dreams. Maybe that's why it's such a popular style today. It helps keep our spirits up.

Meanwhile in Pennsylvania, the great American architect Frank Lloyd Wright was designing his own version of a mid-century modern kitchen.

By the late 1950s, Wright was obsessed with exploring images of the space age in his projects: a swooping, curvy flying-saucer church, a sleek shard-of-glass skyscraper, elaborate domes, twinkling needle-like spires, plans for a futuristic city where people would travel by flying saucer.

I find there is something incredibly pleasing about being in a Frank Lloyd Wright house. Whether it is the sweeping horizontal rooflines, muted light, rich natural materials or glittering art glass windows, he never disappoints. It is the case with every Wright house I have visited. He is my favourite architect.

For years, I have had a fantasy of cooking in a Frank Lloyd Wright kitchen. It is because when I tour Wright buildings, I am rarely allowed to touch any of the furniture, sit in the chairs or even lean against the walls. Everything is so very precious. But I love kitchens and I like to get my hands right into the cookie dough.

My fantasy includes making a baked Alaska in a Wright kitchen. Baked Alaska is a dessert that was all the rage in the late 1950s. Although it dates to the nineteenth century, it made a comeback when Alaska became a state in 1959. Cake, ice cream, meringue and fire—what's not to like? It also has that domed flying saucer look to it—a perfect mid-century-modern-shaped dessert.

But baked Alaska is back again. The popularity of mid-century style has inspired retro parties with baked Alaska as the feature dessert. Recently, Franny and I were on a boat cruise, and guess which dessert was the finale to dinner on the last night? The lights went down and, with great flourish, a team of servers strode out carrying flaming baked Alaskas on silver platters above their heads. Oohs, aahs and joyous applause filled the dining room. For me, it's a treat that has never been out of style.

A house called Kentuck Knob was one of the last Frank Lloyd Wright designed—and it has an extraordinary mid-century modern kitchen. Kentuck Knob is sometimes overshadowed by its better-known neighbour, Fallingwater, a spectacular Wright residence built over a waterfall. Nevertheless, I am focused on a visit to the more modest Kentuck Knob because it will result in all kinds of new ideas for our perfect kitchen.

When I called the staff at Kentuck Knob and explained that I wanted to make a baked Alaska in their Frank Lloyd Wright kitchen, I expected a big fat no (as in "Are you *crazy*?"). But they said yes! They were really excited by the idea. I couldn't believe it. I packed up my pots and pans, electric beater and dome-shaped bowl and got in the car right way.

The narrow roads to Kentuck Knob swoop up and down the steep Allegheny Mountains and are bordered with lush green forests and jagged rock outcroppings. I take a few wrong turns but make a magical stop to say hello to a mother doe and her baby munching grass at the side of the road.

After about an hour's drive south of Pittsburgh, I pull up to the Kentuck Knob Visitor Center, where I meet a friendly woman named Emily Butler who is the head of preservation and conservation for the house. Emily and I shake hands and I tell her I can hardly wait to see the kitchen.

"We are expecting you, Mr. Ota," she says with a big smile. "We're looking forward to eating baked Alaska."

Emily tells me she is currently doing her master's in historic preservation at Tulane University and writing her thesis on preserving Frank Lloyd Wright houses. I am glad to know that I am with a fellow Wright fan. The day is off to a great start.

In preparation, besides my shopping bag full of kitchen equipment, I have packed a dozen eggs, cartons of Hagan ice cream and a flask of rum for the flambé finale. I grab my bags, and Emily and I drive up a winding forested hill toward the house.

Kentuck Knob does not have the simple horizontal lines of a Wright prairie-style house. It was finished in 1956, and by that time Wright had long moved on to his futuristic mid-century modern style, leaving the prairie way behind. The house spreads out before me. I am breathless at the sight of a series of massive copper roofs that float above a low-lying house. Sleek and aerodynamic, the house has the lines of a supersonic stealth bomber about to take off.

Kentuck Knob was conceived when Isaac and Bernadine Hagan of Uniontown, Pennsylvania, visited Fallingwater. The Hagans, who owned an ice cream company, loved the house so much that they commissioned Wright to build a similar residence for them. At that time Wright was eighty-six years old and hard at work on the Guggenheim Museum in New York. It was to be one of the last homes he completed before his death in 1959.

We walk around the outside of the house, and I can see that Wright has designed the angular structure to wrap around a wide courtyard. The one-storey, two-thousand-square-foot residence is nestled into the hillside, facing south to best soak up the rays of the sun. Wright had the walls built from fieldstone found on the site so that Kentuck Knob appears to be part of the mountain itself.

The building process did not go as smooth as silk. The Hagans set a construction budget of $60,000, but Wright went over budget and the house ended up costing $96,057 to build. In addition, there were structural problems and delays. The roof system and cantilevers required redesign and reinforcement.

On my way to the double glass doors at the entrance, I spy a bright red tile with Wright's signature that is embedded in the wall, indicating a Frank Lloyd Wright home design. I touch it for good luck. But then Emily gives me a smile and tells me that Wright, on top of going over budget, had the audacity to charge the Hagans an extra $2,000 to implant the stone into the wall.

Wait a minute! Did I just detect an affectionate little jab at Mr. Wright?

As I walk through the doors, I immediately feel the serenity of being in a Wright house. The first thing I see is a massive fireplace of angular stone walls that anchors the entire interior of the house. I walk past it and am drawn into the living room by natural light flooding in from a band of south-facing windows. Open and intimate, the interior of Kentuck Knob is a composition of stone, wood, glass and natural sunlight. The stone walls of the outside continue inside, as does the stone floor. Wright seamlessly integrates the interior of the house with the outside.

I find that the most outstanding aspect of Wright's architecture is his commitment to nature. He is able to create beauty and wonderment with every view across a terrace, every red cypress wood connection and every pane of stained glass. Kentuck Knob is an excellent example of Wright making the house one with its surroundings.

As I look closer, I realize that Wright has designed the house in a series of hexagons. The floor plan, furniture and fireplace are hexagons. I look outside and even the skylights on the terrace overhang are in the shape of hexagons. There are almost no right angles, adding to the wonder of the house.

"Wright forbade art on the walls," says Emily. "He felt it would take away from his architecture."

Oh my goodness, it's true. Emily is gently teasing Mr. Wright. Wright was not the most modest or flexible architect in the world, which adds to his mystique and colourful character. At many Wright sites, the guides speak of him as if he is a god or a genius and never say anything negative. I find Emily's commentary at Kentuck Knob to be refreshingly balanced about describing his tremendous architectural talent and his personal idiosyncrasies.

Wright has designed the house in a series of hexagons. There are almost no right angles, adding to the wonder of the house.

Height mattered at Kentuck Knob. As I walk through the house, I feel that the ceiling is unusually low. At first I think I am imagining things, but no. Wright was five-foot-eight, a height he considered the human ideal. (Some say he was actually five-six, but that cannot be substantiated.) And so he designed the proportions, shelves and ceiling heights of the house according to his own stature. Unfortunately, the owners had a son who was six-two, and the Hagans requested that the ceilings be raised to accommodate him. Reluctantly, Wright compromised, raising them to six-foot-seven. ("Oh, all right!" I can just hear him saying.)

Part of the living room is glassed in to enclose an angular dining room. Light pours into the dining room through a series of expansive windows that also pulls views of the distant Allegheny Mountains right into the house.

As I stand being warmed by the sun in the dining area, I know that the kitchen can't be far away. I wander through a narrow opening in a stone wall toward the middle of the house and suddenly there I am in the kitchen. I am surprised, because in contrast to many houses of its time, there is no door to separate the kitchen from the dining area. Instead, Wright has partially hidden the kitchen from the dining room by angling the stone wall to block any view of Mrs. Hagan's cooking activities.

Wright has made a statement about the importance of the kitchen by locating it at the centre of the house, in between the living and bedroom wings. Enclosed in a massive stone tower, the kitchen is expressed on the exterior as well as the interior. It reflects the Hagan family's love for cooking and entertaining and was the hub of domestic activity. There were no servants at Kentuck Knob. Mrs. Hagan had a kitchen she could cook in.

Although Wright was recognized in 1991 by the American Institute of Architects as the "greatest architect of all time," his kitchens are sometimes a disappointment. Compared with the dazzling art glass windows, the dramatic fireplaces and his artistic use of stone and copper elsewhere in his houses, the kitchens can be surprisingly ordinary.

An exception is Kentuck Knob. Here, the kitchen is the architectural and spiritual heart of the home.

The first thing I notice is that it is a large kitchen by Wright standards. Towering stone walls rise fifteen feet and form a hexagon-shaped room. Wright designed the room with no windows, so the only sunlight comes from a hexagonal domed skylight in the ceiling. It creates an inspirational glow. It reminds me of the soft light over a church altar.

The skylight was originally a clear dome, but Mrs. Hagan had translucent shading and a grid installed to diffuse light and heat. This was much to the objection of Wright, who felt her changes compromised his design.

As I look around the room, I begin to take in more of Wright's design. Red cypress cupboards mounted on the stone walls match the cabinetry in the rest of the house. The honey-yellow stain applied to the plain cupboard doors allows the natural wood grain to come through. I notice that the cupboard doors are mounted with piano hinges that run the length of the door to provide extra strength and prevent warpage. Hand-carved duck decoys and ceramic pots sit on open shelves, a tribute to the Allegheny outdoors.

However, that is where the rustic ends. The rest of the kitchen is space-age modern Frank Lloyd Wright style. Wright and Mrs. Hagan went all out with the shiny new-age look. Kitchen counters are cut at angles to echo the geometric vibe in the rest of the house. Mrs. Hagan chose to cover her countertops in chic stainless steel, which gives them a gleaming profile and pulls all the kitchen elements together. (Stainless steel was also the material used for countertops in the Hagan ice cream factory.)

The original 1956 Westinghouse oven is similarly faced with a stainless steel door. The appliances are built-ins, a popular style choice in mid-century modern kitchens. The clock and the temperature-control dial are housed in two circular windows under the oven door and look as though they belong in a jet fighter cockpit. A sleek stainless steel exhaust hood mounted under the cupboards has an understated horizontal profile that adds to the streamlined look of the kitchen. I am pleased to feel the cook-friendly cork floors under my feet, another new mid-century modern material and a Mrs. Hagan selection.

But the fold-back stove burners are the big WOW in this kitchen. They are the equivalent of a disappearing stove. The mid-century modern look dictated a clutter-free, pristine kitchen. To achieve this aesthetic, appliances were designed to fold away and disappear. Kentuck Knob's space-saving "Fold-Back" burners were made by Frigidaire with island and galley kitchens in mind. They gave the mid-twentieth-century homemaker extra counter space for food prep and had a no-fuss appearance when folded away.

Each pair of burners sits in its own mini stainless steel case. When they're in the up position, their stainless steel bottoms give the kitchen extra shine—you don't even know they're there. In true '50s style, the fold-back units are also detachable from the wall, so they could be taken outside and used on the patio. The heat is controlled by a knob called the "Heat-Minder." The control panel has six settings—Med Hi, Med Lo, Low, Simmer, High and Off. I am amazed that after almost seven decades of service, these burners still work. The retro "F" Frigidaire logo with a gold crown has reason to sit proud on the control panel.

The burners are adorable, and I wish we could have a couple of these for our perfect kitchen. It's almost like having a pet stove.

In 1956, the Kentuck Knob kitchen must have been cutting-edge modern. In fact, it's still astonishing to me today. It is my favourite kitchen of all the Wright kitchens I have been in.

But what was it like for Mrs. Hagan to cook here? I am about to find out.

I am in trouble.

Gazing around the Kentuck Knob kitchen, I find myself in a mystical architectural trance. I know I am here to cook, but I can't snap out of it. All I can do is look around the kitchen in awe. I open and close the oven door. I run my hands across the countertops. I am lost. I am overwhelmed by where I am. I know that I must start to cook, but I don't know how. My mind is a blank.

I reflect that it might have been a good idea to practise making a baked Alaska before coming to Kentuck Knob. That would have made a lot of sense, considering the complexity of the preparation, the distance I have travelled from Toronto and the trouble I have put the Kentuck Knob staff to. Most sane people would do that. Yet I have come to this Frank Lloyd Wright kitchen having never made a baked Alaska. Ever. I am winging it today.

And now I stand in the middle of the kitchen paralyzed.

Thank goodness Emily Butler comes to my aid. She can detect my anxiety attack and she starts to remove the contents of my bag . . . cake pans, electric beater, bowls, spoons. She pulls out my cookbook and opens it up to the recipe for baked Alaska. But I am so panic-stricken, the words don't register. I truly am a mess.

Emily advises me to lay out the sponge cake I had baked the night before. I fumble about but do manage to follow her instructions.

She checks the recipe and pronounces that the next step is to build the ice cream filling. She directs me to line my metal bowl with plastic wrap. I do that, then begin to press layers of ice cream into the bowl.

I am using Hagan ice cream, in honour of Kentuck Knob's first owners, and am alternating layers of vanilla, green mint and brown chocolate to mirror the colours of the clouds, surrounding trees and Allegheny rock outcroppings. I hope Mr. Wright would approve. I pack the ice cream into the bowl with so much nervous energy that it's rock solid.

I feel myself coming out of my stupor. I'm getting warmed up.

The next step is to separate the whites of four eggs by breaking each one open and transferring the yolk from shell to shell, letting the whites fall into a metal bowl. I add sugar, plug in my trusty electric beater and whiz the egg whites until they form stiff peaks.

I feel myself picking up speed. I flip the round cake onto a cookie sheet. The cake will be the base for the dessert. The packed ice cream drops out of the bowl on top of the cake. I spread my meringue over the entire ice cream dome with a knife.

"Make sure your meringue completely covers the cake layer right down to the edge of the pan," says Emily with urgency. "That way, the ice cream won't melt because it's insulated by the meringue. The dessert is cooked directly under the hot broiler. As long as the ice cream is completely enveloped by the meringue, it will stay cold."

I finish frosting the ice cream dome with meringue and step back. Emily and I are quietly impressed. The concoction looks like a huge snowball. I pick up the cookie sheet with the dome on it and carry it over to the 1956 Westinghouse oven ready to slide under the broiler.

But wait a second. I don't even know if the broiler works. I take a look at the front of the oven. The circular clock is broken, but miraculously, the round oven dial is operational.

My panic returns. If there is no broiler, the whole idea of a baked Alaska is down the drain. The broiler is crucial to toast the meringue. I hadn't even thought to test the broiler.

"Does this broiler work?" I ask Emily with a pleading, hopeful tone to my voice.

"I'm not sure," she replies. "We never use it."

"You mean the broiler hasn't been turned on since 1956?!"

We crouch around the open oven door. I turn the dial all the way around to "Broil" and hold my breath. For a few seconds nothing happens, and I get ready to declare my trip a complete failure. But suddenly I detect a musty smell in the air, and a wisp of smoke floats from the roof of the oven. I stretch my arm into the oven and it feels warm. I look at the top of the oven and it's turning bright orange.

"Hallelujah!" I yell. "It still works!"

Sixty-odd years later, we are still in business. We pop the domed snowball inside the oven and close the door.

Emily and I look at each other and then begin to laugh. I feel like I'm in a Charlie Chaplin movie.

After a few minutes, the top of the dome is turning a toasty golden brown, but the sides are still white. To compensate, I leave the baked Alaska in the oven a bit longer, but the top starts to turn black. I have to get my baked Alaska out of the oven *tout de suite*.

"Quick," I say to Emily as I watch the smoke thicken. "I need the oven gloves."

"What oven gloves?" she says.

"Oh no!"

In a flash, we are both opening and closing the red cypress kitchen drawers, frantically looking for oven gloves or a reasonable facsimile. Thank goodness Emily finds some gloves in a lower drawer, and I gently slide the pan of smoking baked Alaska out from the oven. I am careful to avoid spilling any melted ice cream onto the bottom of the oven.

Now that the slightly singed dome is out of the oven, it is time for the rum flambé. My plan is to warm the rum in a saucepan over the 1956 flip-back burners. But my frustration reaches a peak when I realize that I have forgotten to bring a small saucepan. This completely breaks my heart because I wanted to cook over those flip-back burners. Again, Emily searches through the kitchen cupboards and finds a perfect saucepan. She is amazing.

I pour a little rum into the saucepan and turn the Heat-Minder to medium-low. After about two minutes over the burner, I strike a match and light the warm rum in the saucepan. We watch as the alcohol catches fire and a tranquil blue flame peeks over the top of the saucepan.

The kitchen gods seem to be smiling on us now, and before they can change their minds, I pour the flaming rum over the baked Alaska. The blue flame gently dances and flutters and spreads over the entire dome. It is captivating.

We step back and admire. Except for a few small crusty black patches on the top, which we ignore, the baked Alaska looks almost perfect. The flaming dome is a thing to behold. It is a glowing architectonic masterpiece. I turn the cookie sheet around to hide the burnt spots. I think of Franny and our baked Alaska dessert on the river cruise. If I blur my eyes a bit, this baked Alaska looks like it could also be served on a cruise ship. Well, almost.

Somehow in the excitement I manage to slice and serve.

Emily and I dig in at the same time and we are in equal parts ecstasy and shock. The baked Alaska is amazingly delicious. The meringue is toasted golden brown, and when I crunch through the shell and spoon some into my mouth, it puffs into egg white happiness not unlike the velvety lightness of a floating cumulus

cloud. I can feel the hot and cold sensations of the dessert on my tongue at the same time. It brings to mind the ice cream profiteroles at Monticello.

In between the luscious meringue and moist cake base are hits of vanilla, mint and chocolate ice cream that meld together into a creamy sweet confection that coats my mouth with pleasure. The flambé proves to be more than a visual wow, as it leaves a pleasing rich rum aftertaste that finishes off the whole culinary experience. I swirl every one of the sweet flavours and textures around in my mouth. I zero in for another bite.

Emily calls the Kentuck Knob staff and groundskeepers to join the party. There is enough for everyone. There are collective moans of pleasure. They can't believe baked Alaska can taste so good. They remark that the cross-section view of layered cake, ice cream and meringue is visually impressive. During a time when looks counted, I can see why this was such a popular dessert.

My adventure is a success. My dessert is not perfect—it could have been a little less burnt—but cooking it in a Frank Lloyd Wright kitchen has been great fun. There is much cheering and hugging in the kitchen. It is a Kentuck Knob team effort.

We look skyward and toast the Hagans and Frank Lloyd Wright with forkfuls of baked Alaska and big smiles. The quintessential mid-century modern dessert in the quintessential mid-century modern kitchen. I hope they are looking down at us.

And I'm sure Mr. Wright is rolling his eyeballs.

As I pack up to leave Kentuck Knob, I reflect that this is a house that Franny and I could easily live in. While Fallingwater is a spectacular show house built over a gorgeous waterfall, it is filled with too much daring and audacity. It would be like being married to a supermodel.

Kentuck Knob, with its functional kitchen, comfortable spaces and quieter approach to its environment, would be a calmer fit for us. It is one of my all-time favourite Wright houses.

The Hagans lived at Kentuck Knob for almost thirty years. They loved the house as a place to bring up a family as well as for cooking and entertaining. In 1986 the family sold the house to Lord Peter Palumbo of London, England, who opened it to public tours in 1996. The Palumbo family adore the house and live in it for several months each year.

Thanks to the Palumbos and the Kentuck Knob staff, baking a baked Alaska in the Wright kitchen is one of the Top 10 Most Fun Things I have ever done in my life. It was a roller-coaster ride of panic, confusion, exhilaration and amazement all stacked up on top of each other like my layers of ice cream.

My dessert may not have been perfect, but as I walk past the $2,000 red Frank Lloyd Wright signature stone in the wall, it occurs to me that neither was Frank Lloyd Wright.

Kentuck Knob
723 Kentuck Road, Chalk Hill, Pennsylvania
www.kentuckknob.com | (724) 329-1901

Dear Franny,

I love the optimism and hope that is inherent in the mid-century modern style. There are times when hope is a rare commodity. It would be nice to find it in our kitchen. Here are some ideas that I discovered at Kentuck Knob for us.

We can do our Wright best to bring nature into the kitchen—plants, stone and wood. It can only inspire health and well-being in our lives. I loved the way Mr. Wright blended natural materials from the site with mid-century modern steel and innovation.

Mrs. Hagan loved her kitchen and she had ways to make cooking easier. Everything was within two feet of her. We should have a "go-to" counter space for chopping and prep work where spices, utensils and appliances are all within easy reach.

The Hagans knew the positives of stainless steel counters from their dairy—easy wipe-down, cleaning and sanitation. Wright had wanted Cherokee-red tiles, but the Hagans went with stainless steel.

The cork on the floor is a natural material, easy on the feet, looks good and I'll break fewer glasses.

Flip-back burners are off the market, but any idea to keep the counters clean and clear is worth exploring.

I know you like baked Alaska. My version is a little crispy on top.

XOXO

J.

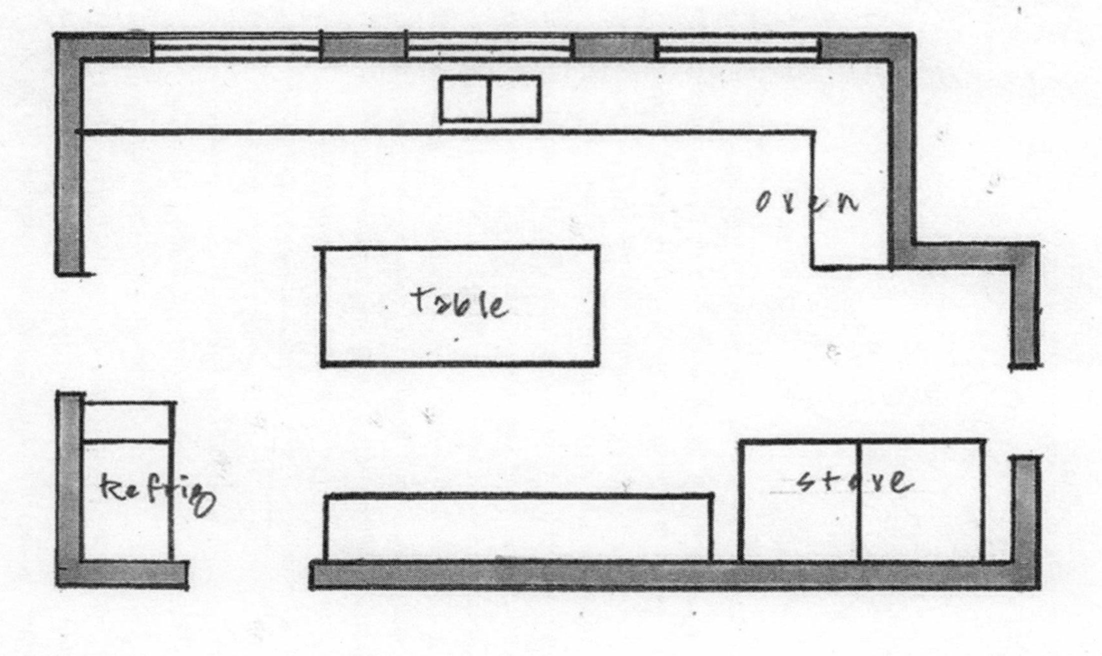

Plan Julia Child Kitchen

Interior Julia Child Kitchen

10

JULIA CHILD KITCHEN, 1961

Washington, D.C.

Even though she's gone, I'm still afraid of my mom.

When she died, it left a big hole in the lives of our family and her friends. She was a live wire—fun and always the life of the party. Whether she was rolling sushi in the middle of the night, decorating a Charlie Brown Christmas tree with homemade ornaments or pitching a tent in the rain, she tackled everything with gusto, energy and an enormous sense of humour that was the trademark of her life. And she loved to cook. Kishka, falafels, osso buco, ha gow, siu mai, somen noodles . . . she did it all.

One consolation in my grief was leafing through her extensive collection of cookbooks. Especially those written by Julia Child. My mom was an enormous Julia Child fan and watched her at every opportunity. I remember Mom parked in front of the TV, wide-eyed, with an enormous grin on her face. "Look at that Julia," she would say, beaming with admiration. "Isn't she great? Isn't she wonderful? And she's *so* funny." My mom especially liked the shows when Julia would do something outrageous, like suddenly pull out a blowtorch and don a huge welder's mask to brown a hard sugar crust on a crème brûlée.

Julia Child, whisking, basting, tasting, mixing, holding a vanilla bean up to her nose to breathe in the aroma, hoisting up a thirty-six-inch salmon for the camera, instructing viewers on how to roll a party-size bûche in her falsetto voice, but most of all smiling, sharing and having a great time cooking. Her goal was to expand the cooking horizons of North Americans, and my mom in Toronto was with her all the way.

When I picked up each of my mom's Julia Child books, the pages would automatically flop open to favourite recipes. One of them, however, was so full of favourites it was never going to fall open again. It was falling *apart*. My mom's copy of *The French Chef Cookbook*—published in 1968, priced at $2.25 and boasting "Over 1,000,000 copies in print"—is now tattered, spineless and held together with elastic bands.

The most stain-covered page carries the recipe for Soupe à l'Oignon Gratinée (Onion Soup Gratinéed with Cheese). Julia writes of the added cheese, "This turns onion soup into a hearty main course; All you need to complete the meal is a bottle of red wine, perhaps a green salad, and fresh fruit." And this is exactly what my mom did. This was her "knock it out of the park" recipe.

My mom's guests who were invited to lunch for the first time probably assumed they would be served sandwiches, as was the convention at the time. But instead they would be surprised by steaming bowls of French onion soup smothered in melted Swiss and Parmesan cheeses. I was amazed when, at every lunch, people devoured the rich brown soup with the gooey cheese that stretched over their spoons, not to mention the tossed green salad and crusty baguette. Soupe à l'Oignon Gratinée became my mom's signature dish. Repeat lunch guests would look forward to having it again and again. Of course, it also helped that my mom followed Julia's advice and continually refilled their goblets with red wine.

Dessert would be fresh strawberries, just as Julia advised in her book. Except my mom made a slight variation, spooning the strawberries over vanilla ice cream and then drowning it all in Grand Marnier. I like to think that Julia would have approved. The sweet, orangey liqueur was a crowd-pleaser, and Mom's dessert was visually magnificent served in cut crystal dishes.

Lunch was always fabulous, and following the example of Julia Child gave us a false sense of sophistication. However, I do remember one of my stodgy uncles not being impressed and asking in a dour tone, "What is this stuff on top of my soup?"

If my mom knew that I was writing about kitchens, she would insist that my quest for the perfect kitchen include Julia Child's. She would *insist*.

No probs, Mom. That's where we're going. Julia is my hero too.

Given her early life, it is hard to believe that Julia Child would become America's most famous chef. Born in 1912, in Pasadena, California, she grew up in a wealthy family and lived a privileged childhood. Early in her life, her spirited, adventurous manner seemed to work against her in a series of unsuccessful journalism jobs. But in 1946 she married Paul Child, whom she had met while they were both working for the Secret Intelligence

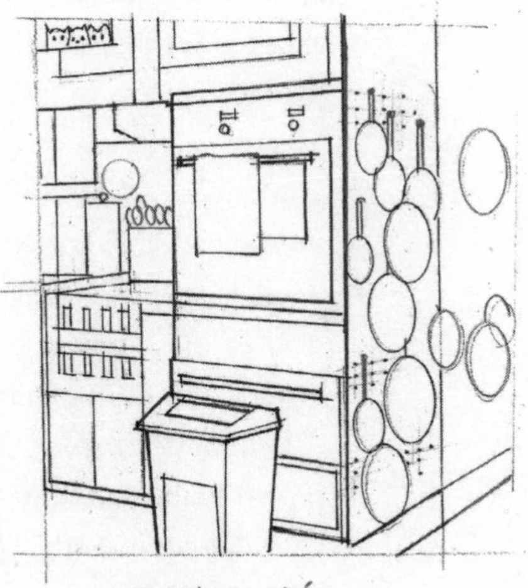

unit at the Office of Strategic Services. They moved to Paris, where Paul encouraged Julia to attend the famous Cordon Bleu cooking school. Julia found her niche in French cooking.

In 1961, Child published *Mastering the Art of French Cooking*. Writing it in collaboration with her friends Simone Beck and Louisette Bertholle, her goal was to make French recipes accessible to Americans, encourage them to prepare food with fresh ingredients and expose them to the pleasures of fine French cuisine. It was an ambitious enterprise. Julia Child was swimming against a tidal wave of instant food.

Food corporations at the time were devoting all their expertise to concocting new processed foods for mass consumption. I present a few examples . . .

My Top 5 Dubious Food Innovations of the 1960s:

#5. Instant non-dairy creamer: A milk substitute that required no refrigeration. Quick, convenient. No real milk necessary.

#4. Powdered dessert topping: Replacement for whipped cream that could be stored in the cupboard. Quick, convenient. No real cream necessary.

#3. Powdered orange drink: Developed for astronauts in outer space. Earthlings could drink this substitute even though we had easy access to real oranges. Quick, convenient. No real oranges necessary.

#2. Instant breakfast: A sweetened powder in an envelope to be mixed with milk that was supposed to replace real breakfast food such as bacon, eggs, oatmeal and pancakes. Quick, convenient. Just add milk.

And my #1 Favourite Dubious Food Innovation of the 1960s:

TV dinners! Frozen chicken and mashed potatoes in foil. Pop into the oven and set up your fold-out TV table. Quick, convenient. Just add TV.

The early 1960s idealized image of a woman was that of a perfect housewife, immaculately coiffed, wearing a crisp tailored dress and pearls, with a house as neat as a pin. The kitchen, situated at the back of the house, was off limits and guests were never allowed in. The kitchen was the woman's kingdom and hers alone.

But some women, like my mom, didn't buy this image. Neither did Julia Child. Happily, Julia found a new ally to market her cookbook, and that ally was television.

A major turning point took place in 1962, when Julia appeared on a Boston TV station with her new cookbook, *Mastering the Art of French Cooking*. When she took the stage, she had the chutzpah to demonstrate how to prepare an omelette on the air. It was the omelette that changed the world. Julia Child was an instant hit, and her appearance led to the television series *The French Chef*, which made her a star throughout America. Her early television shows were filmed on sets at WGBH Boston, but her last three programs, in the 1990s, were shot in her home kitchen, fitted with television cameras, lighting and microphones.

I had read about Julia Child's "little house" at 103 Irving Street in Cambridge, Massachusetts, and even visited it from the outside. Built in 1889, the three-storey residence is a grand grey clapboard house with a tall peaked roof and surrounded by a white picket fence.

And I had learned from our friend David Lillico, a Toronto lawyer and disciple of Julia Child, the present whereabouts of her kitchen.

David is a traveller and collector and the most knowledgeable foodie I know. His long-time devotion to Julia Child extends to owning editions of all her cookbooks, and making what must now be hundreds of her recipes, and, at one time, being a regular at her Toronto cookbook launches.

David celebrates Julia's birthday every August with a dinner party. He selects recipes from, of course, her cookbooks and prepares them himself for his delighted friends. Franny and I were fortunate enough to be invited to one of these sumptuous gatherings, where a toast to Julia Child would start the evening, a Julia dish such as coq au vin would be the main event and a Julia dessert such as tarte tatin would end the feast. During Julia Child's lifetime, David would print up the menu and mail it to her. She would send back a cheerful thank-you note, closing with her trademark *"Bon appétit!"*

David tells me that when Julia Child moved back to her home state of California in 2001, she donated her Cambridge kitchen, in its entirety, complete with cabinets, appliances, cookbooks, table and hundreds of utensils and gadgets, to the Smithsonian Institution. It went on exhibition in Washington, DC, in 2002 and then was taken down and put in storage in December 2011. It reopened in September 2012. And was taken down two weeks later. And *now*, David tells me, the Smithsonian's National Museum of American History has put it up again.

I can hear my mom's voice saying, "Well, John, I guess you're going to Washington."

Washington, D.C., has many magnificent monuments—the Lincoln Memorial, the Jefferson Memorial, the Martin Luther King Jr. Memorial. But my favourite monument is Julia Child's kitchen.

I am standing in line in front of the Smithsonian's National Museum of American History , sweating as the summer sun beats down on me. It has been a long morning, hopping the commuter train into the city and then hiking six blocks from the train station through the monumental Washington streetscape to the museum.

I think my mom would have loved tagging along on this pilgrimage. But I know if she was with me, everything would take a lot longer—more sitting, more slow walking and more snacking on French crullers ("Goddammit, John, forget my diabetes. I need my sugar").

I am hot. I am hungry. But more than anything, I am pumped.

Eventually I make it inside the doors, where a helpful clerk tells me that Julia's kitchen is the opening piece of an exhibition on American food and is on the ground floor, just around the corner from the lobby. Sweat or no sweat, lunch or no lunch, I head straight around the corner in double-quick time.

And when I see it, my jaw drops.

"My god, it looks just like my mom's kitchen."

At first glance, Julia Child's kitchen is mass confusion. The walls are covered with beaters, ladles, strainers and spatulas, all hanging off pegboards. Everything is inexplicably out on the counters, including bottles of vinegars and olive oils, countless pepper grinders stashed into every nook and cranny, dozens and dozens of copper pans, pastry cutters, carving knives, whisks of every size. Even a menacing hacksaw hangs off the pegboard wall, ready to do its duty on a beef rib.

I don't know what to make of it. It is the exact opposite of the white, clean, austere kitchens of today, where everything is hidden away behind cupboard doors. I expected a supersonic kitchen with all the latest gadgets.

But the longer I look, the more it all makes sense. The countless knives and forks are neatly collected in canisters in one corner. A gaggle of measuring cups hangs next to the counter for baking duty. Multiple frying pans hang within reach of the big black stove. The genuine and personal nature of the kitchen is epitomized by the jar of Skippy peanut butter sitting out on the counter. Hey, I'm a Skippy fan too.

Unglamorous—maybe. But basic, homey, happy and completely functional for a person who loves to cook—yes.

The museum's exhibit is designed with vantage points so visitors can look in through the kitchen's windows and doorways. Information panels and video screens of TV episodes explain the history of the room and Julia Child. I notice that as exhausted, bleary-eyed families wander through the museum, they immediately perk up and smile when they see the kitchen. Almost always it is the mothers who pipe up, "Oh look! It's Julia Child's kitchen!"

But as I said—it's also my mom's. The cluttered counters remind me so much of the kitchen I grew up with. My mom had all her utensils and ingredients out and within easy reach while she was cooking. Flour, sugar, jars of knives, forks and spoons, electric can opener, cutting boards piled next to each other, peanut butter, jam, honey, sugar bowl, salt shakers, pepper grinders, coffee mugs and mismatched oven gloves—all ready to go. Even the same red-and-white tea towel hangs off Julia's oven door handle.

Julia Child wrote that if she ever had another house, she would do away with the dining room. It was a waste of space.

Child's kitchen was designed by her husband after they moved into their Cambridge home in 1961. It was their ninth kitchen together, so they knew what they wanted for this one.

In the middle of the fourteen-by-twenty-foot kitchen is a modest wood table covered in a yellow Marimekko print tablecloth. Julia Child had the same kitchen table as millions of other Americans. And she and her husband and their guests ate their meals there, not in a fancy dining room. Julia wrote that if she ever had another house, she would do away with the dining room. She said it was a waste of space in her house.

The table seated six guests. Julia's preference for entertaining was no more than six guests. She felt that when cooking for more than six, the quantity of food became more important than the quality.

The walls and cupboards are painted in a subdued light blue-green. The cupboards and drawers are equipped with simple round metal pulls—the very same pulls that we had in our kitchen. With his wife's height in mind—she stood two inches over six feet—Paul Child designed the kitchen with thirty-eight-inch-high maple countertops, rather than the standard thirty-six inches.

Julia's straightforward light blue cabinets and counters remind me of the Frankfurt Kitchen of 1926. Designed by Margarete Schütte-Lihotzky, Austria's first female architect, it is considered a landmark in domestic architecture and the beginning of the modern built-in kitchen. But the design was not meant to make great food. Rather, the goal was to improve efficiency, hygiene and workflow in the kitchen. Schütte-Lihotzky wanted a kitchen that worked well so that women could spend more time out of it.

The Frankfurt Kitchen was a compact thirteen by seventeen feet, ideal to reduce the number of steps taken by the cook. Clever features included built-in aluminum storage bins for dry goods, a work table with a stool so that the cook could sit while chopping, and a flip-down ironing board in front of the work table. And the kitchen was painted blue because researchers at the time believed flies avoided that colour. (I am still checking into that one.)

My eyes travel beyond the blue cabinets of the Julia Child kitchen to the far end, which is anchored by a hulk-like black six-burner gas stove and oven. It is her legendary Big Garland, a Garland Model 182 that she heaped with affection over the years and a favourite backdrop in photographs of Julia in the kitchen. I would love to have this big, rugged unit in our own perfect kitchen at home, even though it resembles a small Sherman tank.

Next to Big Garland are top-and-bottom wall ovens, chopping blocks and pots and pans hanging on the walls.

I look to the other end of the kitchen and see an entire wall of shining pots and pans and skillets that hang on pegboard panels. In the corner is a black refrigerator that I think is rather small for such a famous chef. It is plastered with fridge magnets and personal photos. Next to the fridge is the kitchen bookshelf, packed tight with cookbooks, many of them by her. I search for my mom's favourite, *The French Chef Cookbook*, but it's not there. But one of my favourites is: *The Way to Cook*, alongside books by others such as two copies of *Joy of Cooking* (one with 1935 marked on the spine), *Larousse Gastronomique* (well-worn), *The World Atlas of Wine*, *How to Clean Everything* and a couple of volumes for birdwatchers. It's a personal and unpretentious collection.

No lacy curtains for Julia. The kitchen faucet and sink sit under a wall of windows that are covered with utilitarian venetian blinds. Sieves, funnels, juicers and a clock are hung on the wall, while a coffee maker and electric mixer sit out on the counter. A red dish rack holds a coffee mug and drinking glass, as though Julia has just stepped out of the kitchen after washing the dishes.

Opposite the windows is a wall of cupboards, under which are blue canisters of tea and a kitchen timer. Oils and vinegars are close to the stove.

The most dominant feature of the kitchen is the vast number of pots and pans hanging everywhere. As I scan the pots and utensils that are meticulously hung on the pegboard-covered walls, I detect an underlying orderliness to the kitchen. Each pot is hung in its specific place outlined in black marker.

This orderliness is camouflaged by the sheer number of objects in the kitchen. But on closer examination, it is visible right down to the way Julia displayed her vast collection of knives in an orderly fashion over the sink. From tiny paring knives to serrated bread knives to long shark-like carving knives, they are all mounted on magnetic strips according to shape and graduated down by size. I must remember this arrangement for our own perfect kitchen so that I don't have to go madly searching in the drawers when I'm in the throes of carving a leg of lamb or peeking inside a salmon fillet for doneness.

I love the look of shiny copper cookware hanging on the wall. Lids for the pots hang over the stove. But more than just eye candy, copper was the favoured choice of Julia Child for its excellent heat-conducting and heat-distribution qualities. It is the largest collection of copperware I have ever seen. There is also a selection of American cast-iron pots. I reflect that if Georgia O'Keeffe kept her Abiquiu kitchen sparse and monochromatic to allow the architectonic elements of the kitchen be the art, Julia Child has gone the other way and made the cooking instruments be the art of her kitchen. "Since we rejoice in the shape of tools," wrote Julia, "cooking utensils become decorative objects, all carefully orchestrated by Paul from pots and pot lids to skillets, trivets and flan rings."

She could cook in the presence of dinner guests thanks to her organized kitchen and her understanding of how the different work zones related to each other.

Overhead illumination is reinforced by spotlights in the ceiling and swivelling lights on the walls to ensure no shadows fall on work surfaces. Julia was a proponent of cleaning as you go to avoid mess.

It's hard to take everything in with the vast display of kitchen utensils and so much else to see. It's only after an hour and a half that I notice Julia liked cats in her kitchen . . . cat tea cozies on top of the fridge (one ginger, one calico), cat fridge magnets and a colourful painting of three kittens peaking over the picture frame hanging on a cupboard door. She had a lifelong love of felines.

Julia also hung signs in her kitchen. Crocks sitting over Big Garland are labelled with masking tape that indicate: "SPOONERY," "FORKERY," "SPATS + MISC." (I was always mortified in front of my friends by the presence of masking tape in my childhood kitchen—but here's Julia . . .) An endearing ceramic plaque hangs high on the wall wishing visitors "Bon Appétit."

My mom also had signs in her kitchen. When she was trying to lose weight, she would tape up a big sign over the sink with the word *DIET* in red crayon. Guests always did a double-take at that.

Today, magazine stylists, designers and real estate stagers want to hide everything in the kitchen away and out of sight. "Neat and tidy" is the mantra. It's almost as though clutter is the enemy. I saw a glossy kitchen magazine featuring supermodel kitchens where there was not one human being present in any of the photos, no sign the rooms were actually used. No kids, no spoons, no pots, no pans. The kitchen not as a place to cook, but as a designer fashion statement.

Julia Child's kitchen is cheerful and feels well loved. Most of all, it is the room of a person who loves to cook.

I take one last look. I remember that when my mom was cooking, she was a flurry of egg cracking, onion chopping, pepper grinding, spatula flipping—an energy-charged woman leaning over the stove in a cloud of steam and all around her were the smells and sounds of cooking food. We ate all our meals in the kitchen, at the kitchen table, just like Julia and Paul. If there was a centre of our house, it was the kitchen table.

Julia kept a large plastic garbage bin with a swinging lid in the middle of the kitchen near the oven. Rather an unsightly object for Julia Child to leave out in the open, you might think. Wait . . . Holy cannoli! My mom kept the very same plastic garbage bin with a swinging lid in the middle of her kitchen! I couldn't believe what I was seeing. I used to be thoroughly embarrassed by the garbage bin in our kitchen and I asked Mom multiple times to remove it. She used to say, "It's fine there, John. Just leave it. It works for me."

OK, OK. Now I get it. I take it all back, Mom. Julia Child would have recognized you as a kindred spirit.

I am delighted when David Lillico invites me to his place to cook a recipe together—a Julia Child cheese soufflé! Now that I've seen Julia's kitchen, I am totally inspired to cook one of her recipes.

When I step into David's kitchen, I find myself in a world of warm amber light and oak cabinetry that reflects both his love of cooking and his passion for Arts and Crafts design. His restaurant-quality refrigerator and gas stove are a perfect complement to the clean, no-nonsense lines of the Arts and Crafts style. I look around at his collection of glimmering copper pots and pans hanging in the air and I think of Julia Child's kitchen.

I uncork the bottle of Sancerre I've brought and pour us each a glass. We toast Julia Child and look skyward.

On the counter next to David is what looks like a flower vase but turns out to be an eight-inch round baking dish for soufflés. David has buttered the sides and bottom and dusted them with Parmesan cheese. I stay quiet and keep my eyes open. Today David is the master chef and I am his assistant.

I tell David, "I am embarrassed to admit it, but I've never had a soufflé before."

"Cheese soufflé is a real classic Julia Child recipe," says David. "Nobody was making soufflés in North America until Julia did in the early '60s. You're in for a real treat today."

David tells me that a soufflé is a baked egg dish that originated in eighteenth-century France. "Soufflés are made up of two things," he says, "a cream sauce, or béchamel, with the yolks as the base combined with the egg whites beaten to soft peaks. The base provides the flavour and the egg whites give the puffiness to the dish."

Julia Child was such fun and a down-to-earth person, so why is it she embraced finicky French cuisine? What did she love about this food? What does it say about her as a person?

"While our Julia liked to have fun," David explains, "she was also very serious about cooking. She felt that the French had the best-tasting food and she followed their meticulous, step-by-step rules of cooking. There must have been a side to her that liked rules and order. She saw being a French chef as a serious profession."

David carries the soufflé dish over to the stove, where he has gathered a copper saucepan, an industrial mixer, a couple of whisks and wooden spoons, containers of spices, and eggs. A small pot of milk is already warming on a burner. Like a maestro, he pauses and looks over his orchestra. "Here we go," he says, and then the action begins.

David melts butter in the saucepan and then adds an equal amount of flour. I think to myself that this reminds me of the roux from the New Orleans étouffée and Jefferson's chicken fricassee. David's making a béchamel sauce. I stand next to him at the stove as he stirs the mixture for a couple of minutes over a medium heat until it becomes a thick, yellow paste. Then he deftly pours the hot milk into the saucepan and whisks vigorously. David looks up from the pan and tells me that the milk must be hot so that it doesn't stop the flour and butter from cooking. He then tosses generous pinches of paprika, nutmeg, salt and pepper into the pan. I picture Julia Child standing at Big Garland doing the same in her blue kitchen.

David had already separated the eggs. He pours the whites into the bowl of his industrial mixer and flips the switch. I watch as the clear, viscous egg whites begin to whirl and thicken before my eyes.

David is in high gear. He turns again to the saucepan and the kitchen fills with the sound of the tsk-tsk-tsk of him whisking the béchamel combined with the high-pitched hum and clatter of the mixer whipping the egg whites. I can almost feel Julia Child behind us, peeking over David's shoulder and offering gentle encouragement. "Excellent, David. Everything looks marvellous!"

In the middle of this action, David whisks the egg yolks into the hot béchamel. His movements are smooth and well practised. As he whisks, the sauce thickens into a custard. With perfect timing, he hits the off button of the mixer and folds big spoonfuls of stiff egg white into the béchamel. "The idea is to incorporate air into the sauce," he says. To me, the mixture looks like a bowl of soft scrambled eggs.

For the next step, Chef David delicately sprinkles grated Gruyère cheese into the custard while he stirs. The sight and smell is too much for me to bear. My stomach rumbles like an earthquake. David scoops the mixture into his prepared soufflé dish and sprinkles more Parmesan cheese on top. He sets the dish in the preheated oven, shuts the door and immediately turns the heat down a little.

David rubs his hands together. "OK, that's it!" he says with a big smile. "Now we just wait for about thirty minutes." From his calm demeanour and smooth movements, I'd say that David has made this soufflé hundreds of times before. I think he could make it blindfolded.

While the soufflé is in the oven, I begin to make my Caesar salad. As I chop garlic, mix the dressing and grate Parmesan, David tells me that he became hooked on Julia Child when he was a teenager, from the first time he saw her on TV.

"She was a person having a lot of fun cooking," he says. "Her enthusiasm was infectious. I remember her trying to carve a roast chicken with a sword, telling a wacky story about Napoleon having to carve his chicken this way when he ate his dinner during the Battle of Marengo. And I remember her wearing a firefighter's hat, holding a fire extinguisher, in case the cognac flames in the crêpe Suzette pan leapt up and threatened to burn down the house. And I remember the time she tasted the dessert that had just been prepared, and she burst into tears because she was overcome by how perfectly delicious it was."

David tells me that he attended four Julia Child cookbook launches in Toronto and owns signed copies of all her cookbooks. Julia's devotees would bring gifts of food to her signings. "Once I had to tell Julia that I'd baked some cookies for her from a recipe in her new book, but in my excitement at seeing her again had left them on my kitchen counter. She put her hand on mine and said, 'David, I hope you live nearby so you can run back and get them.'"

David suggests a final Julia touch to the Caesar I've been making. He takes an egg and drops it in boiling water. In a minute, he takes the egg out and breaks it into a bowl. He pours the still runny partially cooked egg into the salad, and I toss it together with the dressing, lettuce, Parmesan and croutons. Lastly, he adds a sprinkling of sea salt.

David moves to the oven and removes the soufflé. It has risen to a gorgeous golden-brown dome and puffed two inches over the rim. It is a thing of culinary beauty.

"If you wait too long," David says, "they sometimes collapse because the air that is whipped into the egg whites, which has been heated by the oven, cools—so the soufflé falls. That's why soufflés must be served immediately."

David carries the soufflé to the table and sets it next to the salad. His hands go up in the air like a concert pianist's flourish at the end of a virtuosic passage. Immediately, he plunges two serving spoons, back to back, down into the soufflé and pulls it apart. He spoons it onto our plates. It is a delectable sight, and we sample the soufflé without delay.

It is magnificent. A fluffy cloud of cheese goodness floats into my mouth and caresses my taste buds. My mind and body are completely absorbed in pleasure.

The first thing to hit the palate is the melding of the hot Parmesan and Gruyère cheeses in the puffy eggs. Then I experience a bit of crunch that adds to the deliciousness. As I ooh and aah, David tells me that the feather-light, melt-in-your-mouth cheesy flavour comes from the inside of the soufflé and the crunch is from the Parmesan coating on the outside.

We dig into my Caesar salad. It's amazingly good, thanks in part, of course, to the egg, French sea salt and Parmesan cheese.

I raise my wine glass in triumph. "Here's to David Lillico and Julia Child. Two soufflé-making artists!"

I am in culinary bliss. Julia, where art thou?

Cooking itself doesn't explain the enduring popularity of Julia Child. As I watch videos of *The French Chef*, I can't help but think there is more going on than just chopping garlic and whisking eggs. Of course, American women loved Julia because she could cook. But I think it was also because they wanted to be like Julia Child.

My mom used to say that the most important book she read in her life was *The Feminine Mystique* by Betty Friedan, which described the unhappiness of women confined to being housewives and mothers. Friedan's thesis was that the male-dominated magazine industry of the 1960s perpetuated the image of the happy housewife, or the feminine mystique. Up until she read the book, my mom said that she felt trapped in the '60s mould of conformity and perfection. To be like Julia Child was one of the ways my mom could escape that mould. Cooking was a way to express herself.

Happily for Julia, for my mom and for women across America, of all the tedious jobs that housewives had to perform, cooking was the most creative. Not the laundry, not vacuuming and certainly not ironing. Through cooking, Julia demonstrated to

American women that they didn't have to be perfect. They didn't have to be completely obedient. They could have minds of their own. They could be smart. They could make mistakes and they could be funny covering up their mistakes.

And their kitchens could be cluttered.

For millions of women in the 1960s, including my mom, Julia was an alternative role model. Not conventionally beautiful, not blessed with a silver-toned broadcaster's voice, but totally engaging, and committed to cooking, Julia Child was an example of a woman who was wildly successful socially and financially, just by being herself.

When I finally walked away from Julia's kitchen at the Smithsonian, I had a lump in my throat. In my quest for the perfect kitchen, I had gone looking for Julia Child's kitchen, and I found it. But I found something else too.

I rediscovered my mom's kitchen. Her cookbooks, her recipes, how she entertained and that ridiculous garbage bin in the middle

of the room. I also found her challenges as a wife, expectations of her as a mom and her creative outlet in cooking.

Julia Child's kitchen gave me a better understanding of my mom and of the women of her generation.

I just wish I could talk to her about it.

Julia Child's Kitchen

Kenneth E. Behring Center, National Museum of American History

Constitution Avenue NW (between 12th and 14th Streets), Washington, D.C.

www.si.edu | info@si.edu | (202) 633-1000

Dear Franny,

As you well know, my love, I am a minimalist. But Julia Child's kitchen was a revelation. It has changed my whole attitude about not only kitchens but also the interior design of the whole house. Not everything has to be put away. Not everything has to be perfect. Here are some ideas that I gleaned from Julia's kitchen.

The kitchen would be more practical if we design it for cooking—not for fashion.

Ergonomics matter. Julia's counters were custom designed for her six-foot-two height. We're closer to five. I'd have to cook on a stepstool in her kitchen.

Julia took pride in having an organized kitchen. Utensils and ingredients within easy reach.

Let's eat in the kitchen. Or at least have a place in the kitchen where we can sit at a counter and nosh.

Six-burner stove would pump up creativity! But maybe not a Big Garland.

Pots and pans could be decor. They could add a kitchen-celebration ambience.

Personality in the kitchen: art, mementoes, tchotchkes, bright, cheerful, colourful and I love Marimekko.

Let's paint on the wall: "Cooking is not a chore, but an immense pleasure and a true creative outlet."—Julia Child.

We both mention my mom every day—especially you. When you see something in a store or a restaurant, you say, "Oh, your mom would have loved to have this," or "Your mom would have loved eating that . . ." I know my mom would have loved to have seen Julia's kitchen.

XOXO

Your kitchenmaniac husband,

J.

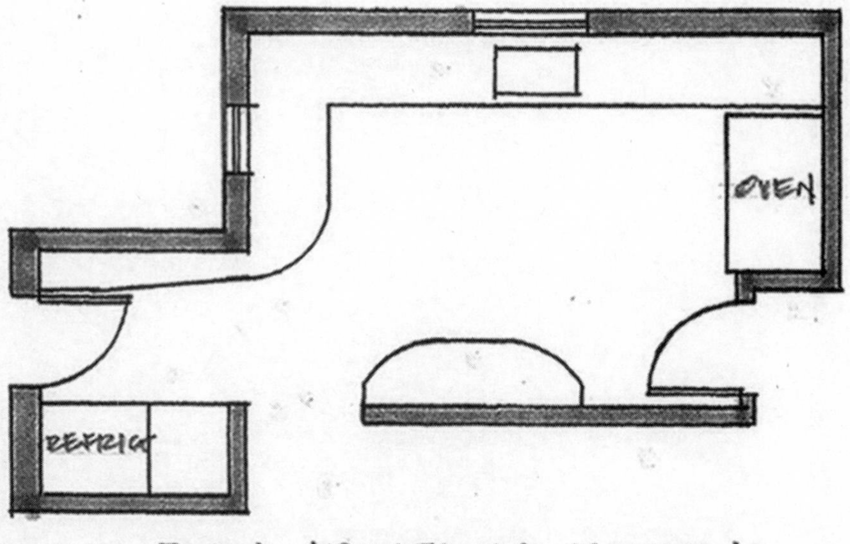

PLAN ARMSTRONG KITCHEN

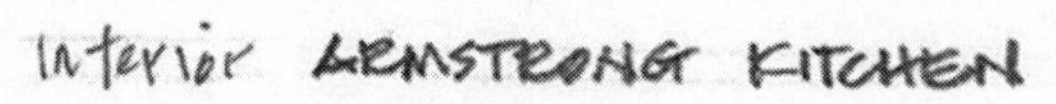

Interior ARMSTRONG KITCHEN

11

LOUIS ARMSTRONG KITCHEN, 1970

Queens, New York City, New York

Jazz was big in our house. The tunes of Ella Fitzgerald, Oscar Peterson, Sarah Vaughan and Louis Armstrong would bop at full blast when I was growing up. My mom had complete control over the stereo when she was making dinner, and every selection came from her own pile of jazz seventy-eights and vinyl albums.

For many, 1964 was the year of the Beatles, but in our house, it was also the year of Louis Armstrong's "Hello, Dolly!" His raspy voice blaring out of our wood cabinet hi-fi speakers brought a smile to everyone's face as he sang, "Hello, Dolly, well, hello, Dolly / It's so nice to have you back where you belong."

The song was inescapable—on television, radio and in record stores. But no matter how many times I heard it, I never tired of singing along with Louis. And I sang it loud and with my eyes open wide like Louis. That was part of the fun.

At the time, I mainly liked Armstrong's cheerful manner. Bulging eyes, smiling face, a gleaming trumpet in one hand and a handkerchief in the other, he always looked like a happy man. Now of course I can appreciate that he was one of the greatest and most influential artists in jazz history.

He also loved to eat.

Knowing about my affection for Louis Armstrong, my friend Katherine Redd at New York's Landmarks Preservation Commission called to tell me about the Louis Armstrong House Museum in Queens, New York. An added bonus for me is that the kitchen is vintage late '60s/early '70s. Those are hard to find. Seventies kitchens tended to be prime candidates for renovation in the next decade. Homeowners got rid of their bright orange and wallpapered kitchens in favour of the new fashion—the all-white kitchen. Efficiency, cleanliness and gleam took over from nature, colour and pattern.

If there's a rare kitchen around, I have to see it. *Especially* if it's associated with Louis Armstrong. And so I'm off to Queens.

I'm in a yellow taxi literally bouncing through a lively Latino section of the Corona neighbourhood of Queens. I tighten my seatbelt as my cab zooms along potholed streets boasting gelato stands, Ecuadoran restaurants and Dominican grocery stores. As low-hanging hydro wires and houses clad in metal siding fly past, I feel I'm in a neighbourhood of proud working people, like my neighbourhood in Toronto.

The driver pulls up in front of a modest two-storey redbrick house with a large bay window. I get out and look up and down the street. An older woman draped in black stares at me for a few minutes before she continues to sweep the sidewalk in front of her house.

I stand there thinking I must be in the wrong place. While I am sure this is a good street to live on, it is not where I would expect a multimillionaire Hollywood movie star and jazz legend to have made his home. It is an unassuming townhouse that blends with all the others on the street. I double-check the address in my notebook. It seems to be the right place.

I peer through the glass doors of a small street-level structure—a former garage—and see people waving to me, beckoning me inside. I see the sign telling me that this is the entrance to the Louis Armstrong House Museum. Yup, it's the right place.

Inside, a smiling man reaches out to shake my hand. This is David Reese, curator of the museum. I explain to him that I have been touring some of the notable kitchens of North America but haven't yet seen an example of 1970s design.

"Well, hold on to your hat," he says.

The museum entrance doubles as the gift shop. Images of Armstrong's smiling face are everywhere. And of course the music playing is "Hello, Dolly!"

David follows me through a door into Armstrong's former recreation room, which now displays examples of his show-business wardrobe as well as honours and mementoes and a gold-plated trumpet given to him by King George V.

His happy and successful life had a difficult beginning. Born in 1901 in New Orleans, Louisiana, Armstrong grew up on a back street in a house with no running water, a dirt floor and little food. The area was so poor that it was nicknamed the Battlefield. When

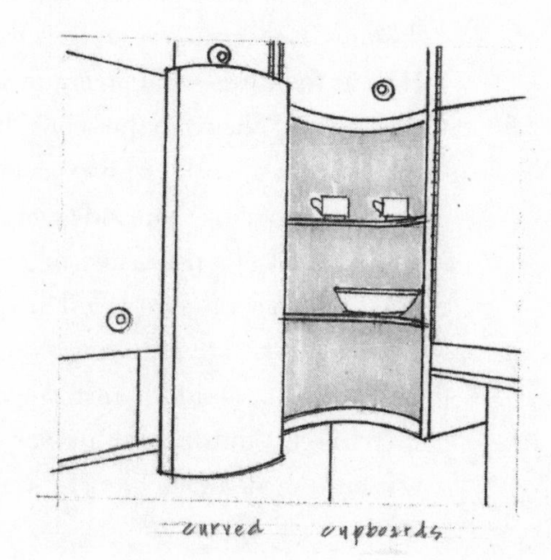

his father abandoned the family, his mother was forced to turn to prostitution. Armstrong was frequently left with his grandmother.

The young Armstrong got a lucky break when the Karnofskys, a Jewish Lithuanian family, gave him a job collecting junk and delivering coal with a pushcart. The pushcart was equipped with a slide whistle (similar to the present-day recorder) on the handlebar, which the seven-year-old Armstrong enjoyed blowing into as he walked the streets of New Orleans—it was his first trumpet.

Although they were poor themselves, the Karnofskys often invited Armstrong to their home for meals. They also, in 1908, loaned him the five-dollar asking price for a coronet he'd spotted in the window of a pawn shop. (He paid them back with a weekly 50 cents.) Armstrong taught himself to play the instrument. The Karnofskys encouraged him also to sing. Armstrong wore the Star of David around his neck for the rest of his life in tribute to the Karnofskys.

Armstrong had left school at age seven to work for a living. A crucial turning point occurred on New Year's Eve in 1912, when he was twelve. He innocently fired his stepfather's gun into the air on a New Orleans street, was arrested on the spot and ended up being sent to the Colored Waifs Home for Boys. It was a blessing in disguise. He was happy at the school and fell in love with music there. He was fed three meals a day, received musical instruction on the coronet and played in the school band. Eventually he became leader of the home's band. Just two years later, in 1914, he left the school as a confident young man and began dreaming of a life making music.

By his late teens, Armstrong was earning a reputation as a fine blues player and played in various bands in New Orleans. But a new style of improvisational music was developing in New Orleans—jazz—and Armstrong entered at the ground floor.

While his musical career was skyrocketing, his personal life was chronically unstable. He was constantly on the road and went through a series of three wives in rapid succession. He met his fourth, Lucille Wilson, in the late 1930s.

Lucille Buchanan Wilson was born in the Bronx in 1914, the youngest of four children and raised in a Catholic home. A musical child, she took piano and dance lessons and eventually went on to dance in the Cotton Club chorus line. It was at the Cotton Club that she met Louis. A passionate romance developed, and Louis proposed marriage.

But Lucille and her parents had reservations about whether Louis was a suitable husband. Never mind his previous three marriages, he was thirteen years older than Lucille and often touring, with all the temptations that a famous man faces on the road. And he did have a wandering eye. But Lucille had fallen in love, and she overcame her apprehensions. After Louis's divorce, they were married in 1942.

Louis never did settle down. He continued to live mostly on the road, by his own rules. He and Lucille clearly reached an understanding. And Lucille—tired of the road and retired from dancing—concentrated on what she wanted: a permanent home, furnished and decorated to the nines according to *her* rules.

She began house hunting. According to Louis's biographer Laurence Bergreen, Harlem was off the list because of all the nightclubs and chorus girls that might tempt Louis when he was home. In 1943, Lucille visited friends of her mother in Corona, a quiet residential neighbourhood. Right next door, a handsome house was for sale. The owners were a white family she had known since she was a child. She had attended school with one of their children, and she knew the house well. She bought it. She didn't feel she needed to consult Louis, who was on tour at the time, as he was for about three hundred days every year. When he finally saw the house, he was delighted with it.

—

David Reese and I walk out to the front of the house and stand on the sidewalk. Before us is a set of brick steps leading up to the front porch.

In the 1940s, Armstrong's income was about half a million dollars a year; he could have afforded a place in a more upscale neighbourhood. But he might not have been welcome. Residential segregation was rampant at the time, with the goal of keeping African Americans out of white neighbourhoods. It was feared that if blacks moved into white neighbourhoods, property values would drop. Banks refused to grant mortgages in or near black areas, realtors would not show blacks houses in white neighbourhoods and the federal government subsidized builders that were creating suburban subdivisions for whites on condition that none of the houses be sold to African Americans.

But Armstrong said, "What the hell do I care about living in a fashionable neighbourhood? The Frigidaire is full of food. What more do we need?"

Another reason Armstrong would have been happy to live (for fifty days a year, at least) in Queens is that it was home to other jazz greats. He could count among his neighbours Ella Fitzgerald, Count Basie, Dizzy Gillespie, Fats Waller, Billie Holiday and Lena Horne.

As David and I walk up the brick steps to the front door, I imagine Lucille and Louis waiting to welcome us at the threshold. I turn the knob on the chiselled wood front door and step inside the house.

The front hall is a small, comfortable space elegantly clad in beige wallpaper embedded with strips of seagrass that feel grainy as I run my hand across it. Dominating the fifteen-by-twenty-foot living room is a portrait of Lucille Armstrong. She is a beautiful woman, with large, captivating brown eyes and a fashionable wavy hairstyle that is set off with sparkling diamond earrings.

Lucille loved to decorate and clearly had a flair for it (as well as the budget and plenty of time). The redecorating reflected the combined tastes of Lucille and Morris Grossberg, an interior designer based on the Upper East Side. Beginning in the 1950s, every renovation in the house, including the kitchen, was jointly designed by Lucille and Grossberg. Louis wasn't involved. All he had to do was come home and be happy in it.

The living room wallpaper is beige and diamond patterned. Two curving sofas and a wing chair are arranged facing each other to make a nook for intimate conversation. On display are Armstrong's mementoes from around the world, including a coral-blue vase from the president of France, African carvings from Ghana and an ornate Bible presented to him by a rabbi in Jerusalem. The white piano against the wall matches the monochromatic colour scheme. Tasteful and comfortable, the room has a subdued mid-century modern look.

A recording of Armstrong on trumpet plays faintly in the background. The museum has incorporated audio effects with Armstrong chatting in his distinctive gravelly voice. The room is filled with Louis and Lucille's laughter and music. The effect is like being at a party where the hosts are just out of sight in the next room. Louis is telling jokes in the hallway and riffing on his trumpet.

I can only guess that beyond the living room is the kitchen. But I sense that David is saving the best for last. We head upstairs.

The spacious master bedroom connects with an ensuite bathroom and dressing room. A glass chandelier with glistening beads hangs from the middle of the ceiling. There's a king-size bed with a white woven headboard. The most noticeable feature, though, is the glitzy mirrored wallpaper with a swirling floral pattern. The bathroom shines even brighter: everything in it is swathed in more mirrored wallpaper—the ceiling, the speakers over the toilet, even the inside of the closets.

Armstrong's office is a tidy wood-panelled den with a reel-to-reel tape recorder system, a turntable built into the wall and several of his jazz LPs lying on his large wood desk. His belongings are everywhere—a typewriter, steamer trunk, books, papers, letters and sheet music. It is his man cave, where he could be by himself or invite friends in to chat, listen to music and relax with marijuana. Armstrong was an enthusiastic marijuana user, which I suppose might have contributed to his happy manner and appetite for comfort foods.

I follow David downstairs and through the first-floor hallway toward my prime objective.

When I step into the twelve-by-fifteen-foot kitchen, I'm surrounded by curved banks of electrifying turquoise-blue cabinets. I can't believe my eyes. I adore turquoise and shout "Yes!" I have had an innate, unexplainable attraction to this colour all my life. I feel as though I am standing in a dreamy aquamarine heaven.

Louis Armstrong was not able to enjoy this dream kitchen for long. The renovation was completed in 1970. Armstrong died the following year. (Lucille continued to live in the house until her death in 1983.)

"Lucille had paint custom-mixed so the cabinets matched her turquoise Cadillac," says David. He opens up the cabinet doors to show me they're mounted on piano hinges to open wide for ease of access. The cabinets have silver pulls and are built to the ceiling to provide maximum storage.

Bright light from the fluorescent panel mounted on the ceiling causes the cabinets and Lucille's early Sub-Zero refrigerator (also turquoise, of course) to glisten.

Turquoise was an excellent choice. The 1970s turned out to be a colourful time in interior design, and Lucille Armstrong was right on trend from the start. The majority of today's kitchens are minimalist layers of white, but when Lucille was renovating her kitchen, it was the complete opposite.

In 1970 turquoise was a hot colour. Turquoise is a sky-blue gemstone mined in the American Southwest. The Navajo, who believe turquoise is a powerful sacred healing stone that connects heaven and earth, started incorporating it into jewellery in the early 1900s, and later Navajo jewellery designers were crafting chunky belt buckles, rings and bracelets. In 1970, the fashion world fell in love with turquoise because it gave visual pop to everything from dresses to jewellery.

David slides open a kitchen drawer to reveal that they contain clever wood slats that fit together like a Chinese puzzle and divide them into quadrants to minimize wasted space. The changeable partitions widen or narrow spaces to allow for organized storage of knives, forks, spoons and spatulas of different sizes. A three-tier Lazy Susan spins in a lower corner of a bottom cupboard for easy access to jars, pots, pans and canned goods. The floor of eggshell-white linoleum has been cut and laid in a custom pattern of diagonal squares.

In my mind I picture Lucille standing over her six-burner stove and double oven with her cookbooks open. Franny has been talking about double ovens for years. She wants to be able to cook several dishes at different temperatures at the same time. I take note.

"Take a look at this," says David as he proudly points to a steel plate mounted on the stove, engraved with the words *Custom made by Crown for Mr. & Mrs. Louis Armstrong*.

Lucille clearly loved wallpaper. Here, it's an umber-and-white floral that covers the ceiling too. It has the effect of giving a cozy feeling to the kitchen. The countertop in white Formica curves around the kitchen and extends out into a desktop for Lucille. Lucille's desk chair cleverly converts into a stepstool with a flip of the seat. Lucille was five feet tall, Louis five-four, and they frequently used the stool to reach high shelves. I am glad to know that Louis and I would see eye to eye.

To reduce visual clutter, an electric can opener and the paper towel holder are concealed in the wall next to the sink, hidden from view by a sliding panel. A real eye-opener for me is the NuTone food processor built right into the countertop, with special holders for its various attachments. I might have to insist on something like this in our perfect kitchen. I once came across a NuTone recipe book with recipes for things like lemon chiffon pie and chocolate malted milk. The booklet boasted, "The kitchen appliance of the future . . . Yours Now."

I am strangely drawn to blenders. Maybe it's the assorted buttons on the panels that I love to push. Or maybe it's the visual stimulation of liquid whizzing around in the glass container. But most likely it's the loud grinding noise of ice being crushed along with various liquids, heralding the arrival of a lovely cocktail. I would guess that the Armstrongs sometimes used their blender for that purpose.

There are also multiple push buttons on Lucille's KitchenAid dishwasher (with "Daily Use" and "Party" settings). Her dishwasher, of course, is turquoise.

The breakfast nook was originally the bedroom when the couple moved in (Lucille's mother lived upstairs). Lucille had the nook converted in the 1960s. It is wallpapered all over in the same umber-and-white pattern as the kitchen.

It is not a particularly big kitchen but it is designed to maximize space and it's cheerful, stylish and exuberant. And—however it looked—it was a real working kitchen in which Lucille loved to cook. Even though the Armstrongs were millionaires, Lucille did all the cooking for them and their guests. She bequeathed all her recipes to the museum archives, along with numerous cookbooks. She seemed particularly enamoured with one from 1962, *The Ebony Cookbook: A Date with a Dish; A Cookbook of American Negro Recipes*. As I examine her copy, I am impressed by the heavily smudged cover and worn spine. It is obvious that Lucille used this book countless times.

Louis Armstrong loved to eat. Next to thrilling audiences with his trumpet, his other great passion was food. While on tour, he ate spaghetti in Italy, sat down to bountiful dinners in Africa and the Middle East, and back in Queens, he had a lifelong love affair with Chinese food, often frequenting the Dragon Seed Restaurant in his neighbourhood. Armstrong even gave his music titles like "Struttin' with Some Barbecue" and "Cornet Chop Suey."

Armstrong also enjoyed being photographed with food, loved talking about dining experiences and saved souvenir menus from his most memorable meals. David Reese shows me photos of Armstrong slurping udon in Japan, judging cooking contests and hosting barbecues in his yard with neighbours. Even though this superstar could have afforded to eat the finest, fanciest cuisine three times a day, it seems it was hard to persuade him to dine on anything but his favourite, red beans and rice. He frequently signed his letters with "Red Beans & Ricely Yours! Louis Armstrong."

According to David, when Armstrong and Lucille were courting in the early 1940s, he asked her if she knew how to cook his favourite dish. Lucille had grown up in New York and wasn't familiar with southern cooking. But she learned and then invited him to meet her parents over a dinner of home-cooked red beans and rice. Louis said it was "just what the doctor ordered." Louis enjoyed the dinner so much, he felt he had to apologize for eating so much. Not long afterward, he asked Lucille to marry him.

If red beans and rice was so much a part of Louis Armstrong's life, then I had to try some. Since the Armstrong kitchen is not operational, I ask David to join me at a New Orleans Creole restaurant called Sugar Freak, on a bustling shopping street in Queens. I've already checked out the menu. Red beans and rice is at the top.

Inside Sugar Freak, people are sipping Dixie beers and soaking up New Orleans jazz. We're greeted by the friendly manager, Franco.

I tell him my story about searching for red beans and rice, and Franco says that he and his restaurant will be delighted to help. David and I settle in at a table and we wait.

As I try to ignore the growling of my stomach, David and I chat Louis Armstrong and New Orleans history. A family tradition in New Orleans was to cook a big dinner on the Sabbath, usually featuring a ham. The cook saved the ham bone to boil up with red beans on Monday, which was also washday. The beans and ham bone, along with vegetables (bell pepper, onion and celery) and spices (thyme, cayenne pepper and bay leaf), simmered all day on the stove while the women were busy scrubbing clothes. At the end of the workday, the steaming beans, ham broth and vegetables were served over rice. Monday washdays are now a thing of the past, yet the dish remains very much a part of New Orleans culture.

I have to change the subject. I'm getting crazy hungry. I pull out a card I bought at the museum gift shop. In the black-and-white photograph, Louis is wearing a striped shirt and eating a large bowl of red beans and rice with a spoon. Lucille sits next to him writing a note on a pad. Behind them on the buffet is a birthday cake with candles. It appears he was celebrating with his favourite meal.

Inside the card is a recipe for "Pop's Favorite Dish" by Louis and Lucille Armstrong. The ingredients are kidney beans, salt pork, canned tomato sauce, six small ham hocks, onions, green pepper, dried peppers and garlic.

I am surprised when I read that Pop's recipe simmers for four and a half hours, with an additional one and a half hours for cooking the ham hocks. I picture Lucille chopping up the ham hocks in her blue kitchen. For her, this was an all-day labour of love.

I place the card on the table to help us feel the presence of the Armstrongs.

Manager Franco comes to the table wielding a large tray of platters. Even before I can see the contents, I catch an aroma of salty smokiness. I'm forcibly reminded that I haven't eaten since early morning.

On the platters are plump kidney beans in a thick crimson sauce of green pepper and onions, hunks of succulent smoked sausage, a fluffy bed of long-grain Carolina rice and a side of braised cabbage. It is an enormous amount of food and, despite my hunger, I worry I might not be able to manage my portion.

I take a forkful of the beans, sausage and rice. At once I experience an intense burst of about eight layers of flavour—smoked ham, creamy beans, salt, tomato, garlic, spices, peppers, all evoking past visits to New Orleans. Franco says that instead of using ham hocks, their chef makes the red beans with andouille sausage, shredded pork shoulder and his own special seasonings. I give his dish a big thumbs-up.

Next, Franco sets down a small dish of collard greens. These are large dark leaves like kale or cabbage, boiled and tender and slightly bitter, with some smokiness from small chunks of salted pork. It is a perfect companion to the wonderful comfort food of red beans and rice. We are also served thick slabs of warm cornbread with a nice crust and soft cakey centre. The bread has a slightly sweet, cheesy flavour, and it's perfect for dipping and soaking up the red bean sauce.

I needn't have worried about not finishing my meal.

—

It is the middle of the night and I am tucked into my hotel bed in New York. I have insomnia, which is usual when I am in a strange bed and away from home. Even though it has been a long day of touring and eating, I am wide awake and reflecting on my day at the Louis Armstrong House.

No doubt Louis and Lucille had a good partnership. But it was really Lucille's house. It was her sanctuary.

It is a place that might better be called the *Lucille* Armstrong House. No doubt Louis and Lucille had a good partnership. They owned a house that was important to each of them in different ways. But it was really Lucille's house. It was her sanctuary. It was a house where she could stay with her mother as she aged. I know about that. Franny and I both watched out for our moms as they aged.

Lucille designed a kitchen with custom paint and state-of-the-art appliances and fixtures that fit her specifications. The entire house, from top to bottom, is a reflection of Lucille's creative spirit. Even Louis's den has an artfully casual organization of his books, music and letters that I would not be surprised to learn was the work of Lucille.

And what a great setup for Louis. He loved to eat and he loved to relax. After the long days and nights of being on the road, Louis could return to the spectacular comforts of home to recharge, reconnect with friends and be pampered with red beans and rice.

Gradually, I feel sleep descending on me and I dream of turquoise kitchens. The only sound in my head is *Hello, Dolly, well, hello, Dolly. It's so nice to have you back where you belong . . .*

Louis Armstrong House Museum

34–56 107th Street, Queens, New York City, New York

www.louisarmstronghouse.org | (718) 478-8274

Dear Franny,

I had so much fun visiting Louis and Lucille Armstrong's kitchen. In a funny way I feel like I met them. Walking around their house and kitchen, I felt as though they were still there.

Here are my thoughts on the Armstrongs' super '70s kitchen for us.

The WOW in this kitchen is colour. The turquoise cupboards are the star of the show. Colour could inspire creativity in cooking. And I love that it is the same colour as Lucille's Cadillac. The turquoise had special meaning for her. She stamped her personality into the cupboards and walls. I know you like blue. We could use it in our kitchen and spread it around the house.

Cooking for big dinner parties, we need dual ovens for multiple dishes. Bring on the celebrations! This will make cooking easier for us.

Curving cupboards and countertop—I'd love to have curviness in the kitchen. Lucille's kitchen curves add visual movement and pizzazz. It's a good contrast to the straightness of the walls.

Lucille celebrated with wallpaper—there's no holding back—so '70s! I remember both our moms wallpapered their kitchens in the '70s too. I'm not a wallpaper man, but I'm open to it to honour Lucille, Louis and our moms. Maybe the fridge—that would be different!

Cupboards all the way to the ceiling maximize storage space. We'll need a stepstool like Lucille and Louis had.

The Armstrongs had wood dividers inside the kitchen drawers to make the most of the storage space. We can play our favourite game—I rearrange the spice drawer and then you rearrange it again.

"Red Beans and Ricely Yours!"

XOXO

J.

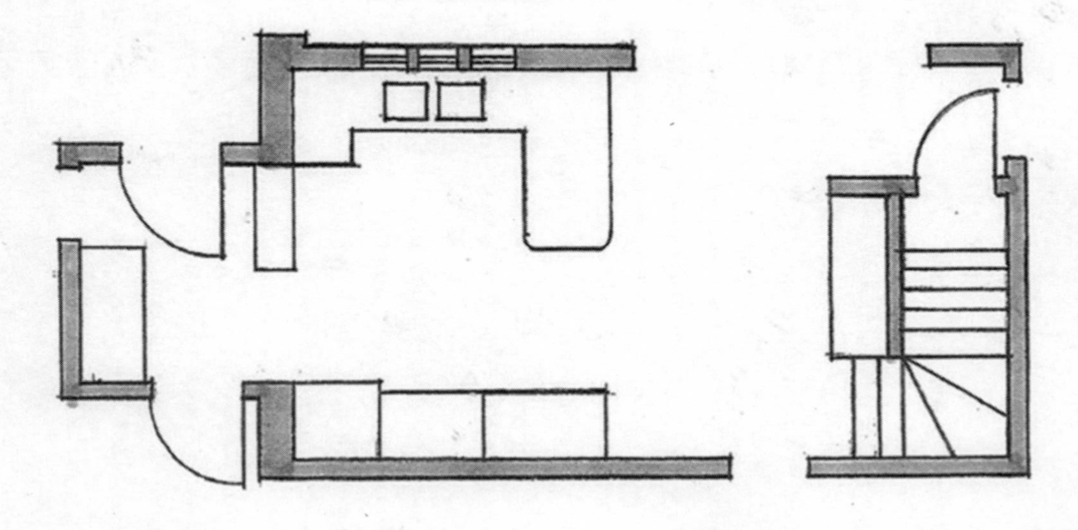

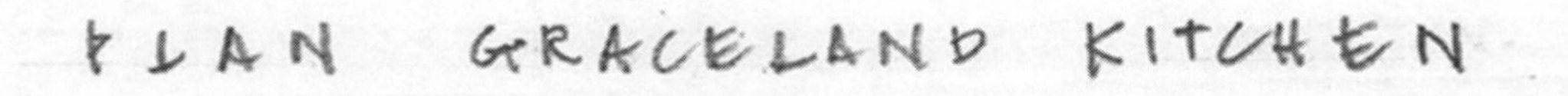
PLAN GRACELAND KITCHEN

interior GRACELAND KITCHEN

12

ELVIS PRESLEY KITCHEN
at Graceland, 1977
Memphis, Tennessee
(and the 1960s Elvis Honeymoon Hideaway Kitchen, Palm Springs, California)

I was all shook up.

On my return home from my happy visit to the Louis and Lucille Armstrong kitchen, I had lunch with my good friend Maria Coletta McLean. As she listened to me describe my Queens adventure, an idea popped into her head. She suggested that I visit Graceland, the mansion of Elvis Presley. She had heard that the kitchen at Graceland was designed in true '70s style. Unrenovated kitchens from that era are rare.

Another '70s kitchen in a musician's house? Fabulous! I am just as much a fan of Elvis as I am of Armstrong. In this case, Elvis will have enjoyed his kitchen a bit longer than Louis did. Both men made great music, had custom kitchens and loved to eat. As part of this kitchen journey, I have to visit Memphis to compare and contrast.

"Wise men say only fools rush in / But I can't help falling in love with you . . ."

Whenever I hear that Elvis song from 1961, I want to drop everything and grab my wife and do a slow dance. And I do—even if it's in the cereal aisle at the supermarket. How does he move me like this? It must be his low, mellow voice, the painfully revealing words, the heartfelt melody. We danced to this song on our honeymoon in Niagara Falls.

Could there be a more unlikely beginning for a person who became an enormous star? Born into a poor family in 1935 in Tupelo, Mississippi, Elvis became, I believe, the most significant cultural icon of the twentieth century. And his music continues to resonate with me.

I am a long-time Elvis fan. How could I write about kitchens and not visit his house, Graceland, and not pay my respects to one of the great music legends? I am not alone in my sentiments. Graceland is the second most visited house in America after the White House.

I had to see the Kitchen of the King . . . *It's now or never*. (Sorry!)

But when I arrive in Memphis and drive toward Graceland, I am a little confused. From a distance, Graceland appears as a quiet two-storey house at the top of a hill, surrounded by oak trees and rolling green pastures. Not how I imagined rock-and-roll royalty would choose to live.

As I walk up the hill and look up at the simple facade of rough-cut limestone, small-paned windows and Corinthian porch columns, I think it is a house more suited for a doctor or lawyer than "the King." I recall a similar feeling when I saw the outside of the Louis and Lucille Armstrong House. The exterior reveals little of the owner's wealth and status.

I take my place in line at the front steps of the house and wait. I read my visitor pamphlet, which offers background about the first owner of Graceland Farms, Stephen Toof, who named the land after his daughter Grace. The land was inherited by Ruth Moore, a niece of Grace Toof. Mrs. Moore and her husband built the house in the Colonial Revival style in 1939. Elvis bought the house in 1957 and moved in with his parents.

As I wait in line to enter Graceland, my mind wanders back to another trip and another Elvis Presley house and kitchen.

A year previously, I had travelled to Palm Springs, California, a city recognized as the world capital of mid-century modern houses. I wanted to find the kitchen of the Elvis Presley Honeymoon Hideaway. As I drove the palm-lined streets, I was breathless at the blocks of 1950s and '60s houses, with their angled butterfly roofs, geometric screened walls and shiny mosaic tiles.

I pulled into a cul-de-sac called Ladera Circle. When I looked up I was astonished to see an aerodynamic house of glass, stone and steel in front of me. An A-frame roof with expansive wings rose above a wide bay window that cantilevered over the street. It floated like a Klingon starship.

On May 1, 1967, Elvis and Priscilla Presley returned from their Vegas wedding and began their honeymoon in this five-thousand-square-foot house. But the house was famous for its architecture even before the coming of Elvis and Priscilla. The 1960 home was designed by architect William Krisel for Palm Springs builder Robert Alexander, who lived there with his wife, Helene. In 1962, *Look* magazine declared it the "House of Tomorrow" for its built-in appliances, innovative furnishings and forward-thinking circular spaces. Tragically, the Alexanders died in a plane crash in 1965.

Later, I met the current owners, who told me, "Elvis loved the house. He thought it looked like a spaceship."

I walked up a series of round concrete steps that wound around a river of flowing water and a lush garden. When I opened the massive front door, the first thing I saw was a sizzling hot sunken living room of glass and stone. It was entirely round. In fact, the whole house was a series of circles.

I stepped down three steps into the living room and toward a stunning sixty-four-foot-long white leather sofa that curved along the rough "peanut brittle" stone wall. I was drawn to a circular free-standing gas fireplace and hovering white vent hood that centred the room. The round living room oozed decadence.

I drifted across the terrazzo floor into the dining area that was furnished with a glass dining table and Parsons chairs. Through the floor-to-ceiling windows, I tried to imagine Elvis, Priscilla and friends lying by the turquoise pool, joking, laughing and sipping cool drinks garnished with tiki parasols.

As I entered the kitchen, my jaw dropped. In all my travels, I had never seen a round kitchen—but here it was. Not only were the walls circular, but everything in the glitzy white kitchen was in circles.

The shiny white cabinets mounted on the walls echoed the shape of the house. The kitchen was relatively small but had a large stainless steel JennAir fridge and double built-in ovens at waist height. The ovens played an important role on the first night of the honeymoon. Elvis had meatloaf and sweet potatoes for his honeymoon dinner.

But for me, the stars of this kitchen were the giant circular stove and range vent. At the centre of the room was a round island with six built-in electric burners. True to form, the burners were arranged in a semicircle, with control knobs in the middle of the island. I pulled back the perforated skirt on the lower half of the island—it folded back like an accordion. Inside was handy storage space for pots and pans.

I was puzzled by a strange-looking black dome on the stovetop. I reached over and pulled up on the handle. Surprise! Under the domed lid was a built-in barbecue, perfect for searing steaks and ribs.

And smoke was whooshed away by a giant range hood that was both futuristic and functional. Dark grey and round, the vent hood took its cues from the lines of a Mercury space capsule. The only thing missing was John Glenn inside.

Elvis and Priscilla must have quite enjoyed their Palm Springs honeymoon house. And the meatloaf did the trick. Exactly nine months later to the day, their daughter Lisa Marie was born.

Back at Graceland, excitement mounts as the line of Elvis fans begins to inch forward. I pass through the front door and all my preconceptions of quiet, conservative house design quickly disappear. I know immediately that I am in the house of Elvis Presley.

Stepping into the entrance hall, I almost have to shade my eyes from the reflective mirrors on the walls and from the white staircase with gold balustrades. The glare of shiny walls, glittering crystal chandelier and gold classical mouldings remind me of the white-and-rhinestone jumpsuits Elvis wore on stage. I notice that there is a colourful stained glass "P" in the transom over the front door.

As I stand in the entrance hall, my Graceland-issued earphones come alive with Elvis singing. "Return to sender, address unknown. No such person, no such zone." I dance by myself. It is great to be at Graceland.

Through an arched opening is the living room, with a white stone fireplace and a fifteen-foot-long white sofa. Photographs of Elvis's parents, Vernon and Gladys, as well as daughter Lisa Marie are on display. Elvis would make visitors wait in the living room and then he would come downstairs from his bedroom to greet them dressed in his white jumpsuit.

Adjoining the living room and behind a doorway framed by large colourful peacocks set in stained glass—very Vegas—is the music room. It is furnished with a black baby grand piano and a 1950s television. I almost expect to see Elvis on *The Ed Sullivan Show*.

The mansion of twenty-three rooms, including eight bedrooms and bathrooms, was decorated by Elvis himself in '70s style. The most unusual room is Elvis's den, known as the jungle room, which is dominated by massive tiki dark wood furniture and features a waterfall gurgling out of a stone wall. The floors, walls and sections of the ceilings are covered in thick green shag rug.

As I pass by the dining room with its shiny black marble floor and gilt table and chairs, I can sense that the kitchen is close by. I don't know what to expect, but this room is the reason I came to Memphis.

Once in the kitchen, I chuckle in disbelief. I feel as though I have travelled fifty years back in time. The kitchen is pure '70s. Unlike the other rooms in the house, which are designed with flash and pizzazz, the kitchen has a homey, unpretentious look and is not overly large.

The first thing that hits me is the vast amount of colour—a collage of wood-panel cabinets, patterned carpet and the ultra-bright colours of the lampshades and appliances. Nothing matches.

Today the preference in kitchens is for minimalist white or a subdued neutral tone. But in the 1970s, kitchens were an explosion of colour and texture. Disco dancing was in, and *Saturday Night Fever* set a fashion trend of colourful pants, shirts and vests that also found their way to interiors.

I designed kitchens during the late '70s, and it was clear to me that my clients were tired of the rectilinear spaces and plain white walls of modern houses. Homeowners wanted to have fun with their interiors. Postmodern architecture was the rage, and people were wanting a change to a warmer, more human design. The Graceland kitchen brings back memories of screeching colour and patterns.

At the centre of the Graceland kitchen is a U-shaped counter of white Formica speckled with gold glitter. Dark wood-faced cabinets with a raised border and steel pulls wrap around the room and provide plenty of storage space. The dark wood gives an old-English-pub look to the space. During his time at Graceland, Elvis had these kitchen cabinets stocked with such items as Pepsi, orange drinks, pickles, brownies, fudge cookies, peanut butter and chewing gum (Spearmint, Doublemint, and Juicy Fruit—three of each).

The glitter-embedded Formica has an affinity with the white jumpsuits worn by the King. By the 1950s and '60s, the heat-resistant and durable properties of Formica, invented in 1910 as an insulator used in electronics and airplanes, made it a popular material for countertops and furniture. At the 1964 World's Fair in New York City, a dream house was built of Formica to showcase its strength and beauty. Visitors to the seven-room, ranch-style bungalow stepped into a domestic world of countertops, cabinets, desks, chairs, dressers and bunk beds made of the laminate fibre and resin material.

The glitter-embedded Formica has an affinity with the white jumpsuits worn by the King.

The range of hues and patterns of Formica expanded in the 1970s so that it became the groovy countertop for the flamboyant kitchens of the period—including the Graceland kitchen. I am delighted to see an avocado-green sink and matching KitchenAid dishwasher sitting under a window that overlooks the jungle room. In the 1970s, interior design responded to a new respect for nature and the environment—and widespread public concern about pollution. It all led to a natural colour palette in kitchens—avocado green, harvest gold and chocolate brown. These earthy colours covered kitchen appliances, cookware, cabinets and countertops.

I am agog at Graceland's wall of appliances. At one end stands a harvest gold refrigerator/freezer (I haven't seen harvest gold appliances in years. It's like being reunited with an old friend), which the Elvis kitchen staff kept stocked with wieners, ground meat, ingredients for meatloaf and its sauce, ice cream—vanilla and chocolate—and at least three bottles of milk and one of half-and-half. Next to the fridge is a stainless steel stovetop with four burners and a griddle. On one of the burners a cast-iron skillet sits ready to make fried peanut butter and banana sandwiches, a favourite of the King.

Built-in double ovens do the job of baking the King's meatloaf, biscuits and roast potatoes. A drawer pulls out from under the ovens to reveal four additional electric burners for extra cooking room. Elvis had a lot of mouths to feed. He was famous for hanging with a male entourage nicknamed the Memphis Mafia who were friends, associates, employees and cousins who served the King. They went where he went.

Near the sink is a stool with a well-worn orange vinyl cushion, a Kenmore electric trash compactor and one of the first microwave ovens available in Memphis. On the counters are a '70s-style push-button telephone, push-button blender, toaster, coffee maker and an avocado-green electric can opener. My mom had the exact same avocado-green electric can opener in our kitchen.

Like many '70s kitchens, the one at Graceland is an assortment of mismatched appliances, furnishings and countertops in different colours and styles. Unlike today's curated white kitchens, it is a cheerful hodgepodge.

Graceland Sink Area

Mismatching is most evident on the floor, with

its wall-to-wall multicoloured carpet in a chaotic pattern of flowers, leaves, triangles, plaids and circles in red, fuchsia, green, maroon and white. The carpet does not match anything in the kitchen. But, of course, that makes it match everything in the kitchen.

I cannot remember when I had last seen wall-to-wall carpet on a kitchen floor, but in the '70s it was cause for kitchen envy. Until the 1950s, most carpets were woven—the carpet produced on a loom the same way woven fabric was—and their high cost made them a luxury. However, in the '50s and '60s, the carpet industry introduced nylon tufted carpet. These had their pile injected through a backing material, so they looked similar to woven wool carpets but were more durable and a lot cheaper. By the 1970s, sales exploded as North Americans embraced wall-to-wall carpeting in a range of piles, colours and textures. Multicoloured carpets in kitchens such as Graceland became popular not just because of their uninhibited spirit but also because of their ability to hide dirt and stains.

Amid all the visual commotion in the kitchen, what brings the biggest smile to my face are three hanging Tiffany-inspired lampshades. Crafted out of stained glass, the shades are designed with a boisterous pattern of colourful lemons, apples and grapes. The light level in the kitchen is low, and these shades act as three beacons, cheerfully brightening sections of the room. They remind me of the paper Tiffany-inspired lampshade that my mom hung over the kitchen table in our '70s house. I went with her when she bought it at a cool hippie-style boutique in Toronto called the Unicorn. It had bright green and blue flower patterns with a black fringe, and she was thrilled to have it in our house. At that time, Tiffany reproductions and copies were wildly popular for their celebration of nature, colour and flower power.

To today's visitor, the Graceland kitchen is a curiosity. But it remains spiritually vital. I admire that people in the '70s used bright colours and had the chutzpah to express their individuality through their kitchen designs. Don't get me wrong—I am not going to consider avocado-green or harvest gold appliances, but somehow, I want to get this uninhibited '70s spirit into our perfect kitchen.

My mom was continually redecorating our kitchen in those years with wallpaper in repeating patterns of loud pink, yellow and orange paisley swirls, painted neon orange cabinets and sci-fi spider plants in macramé plant hangers everywhere. It was a jungle. Above us a giant fluorescent fixture shone down on our orange vinyl chairs and the mismatched patterns on the canisters, saucepans, fondue pots and oven mitts. What was nice was that my dad just let her do what she wanted. He would quietly smile and nod his head.

But the major significance of the '70s kitchen is that it was the beginning of the open concept. Open-plan living first appeared with the Case Study house designs of Charles Eames and Pierre Koenig in California in the 1940s and '50s. Their intention was to bring modern open space to average American homes. But it wasn't until the 1970s that kitchens began to include central islands and incorporate—or extend into—family rooms that gathered the entire family and guests into the kitchen.

The Graceland kitchen is not an open-concept design, but it anticipates it by integrating at one end a small space with a television viewing area. With more open-plan houses starting to appear in the decade, kitchens underwent a major change. It became a more casual space and began to open up to the dining and living areas.

By the 1980s, kitchen designers were painting over the psychedelic wall colours and dark wood cabinets of the '70s with cloud white. But they kept the open kitchen plan that we have today.

In a funny way, I miss that colour craziness. The kitchen was our happy place.

—

I love dinner parties.

I am in Toronto, just a few weeks after my visit to Graceland. Our house is filled with excitement as I plan, shop for and cook a dinner party to honour the King. We have invited some good friends to our house to celebrate. They also happen to be Elvis fans and were very happy to get the invitation.

Elvis could sing but not cook. He left that duty to Mary Jenkins Langston. She had been the Presley family's maid for three years until Priscilla moved in, noted that Langston shared the couple's taste for simple southern fare and promoted her to the position of cook. That was in 1966.

Ms. Langston said of Elvis that "the only thing he got any enjoyment out of was eating." That made the Graceland kitchen one of the busiest rooms in the house. Although the Graceland kitchen appears visually uncoordinated, Langston and a staff of cooks masterfully prepared meals for Elvis, his family, his entourage and guests around the clock.

Elvis kept unusual hours and was known to have breakfast in the evening and lunch at midnight. He was able to order food via an intercom between the kitchen and his second-floor bedroom, a place where he stayed more in his later years. Even with his fame and fortune, Elvis enjoyed eating simple southern fare, the dishes cooked by his mother that he remembered from his childhood: hamburger steak, fried chicken, sweet potato pie, biscuits, banana pudding, and of course the peanut butter and banana sandwiches.

Elvis was indebted to Mary Jenkins Langston for cooking in his kitchen. Over the years, he bought her four cars and a three-bedroom house. Even once he passed away, in 1977, Langston remained cook to the family for twelve more years.

—

For my party, the menu will be pure Elvis, with his favourite dishes that I'll be cooking from Mary Langston's recipes. At the moment I am in the middle of creating an Elvis meatloaf, tonight's main course. I am making this meal in honour of his honeymoon evening in 1967—meatloaf, sweet potatoes and another Elvis favourite, pecan pie. I have set a place at the table for the King with his photo and a bottle of Pepsi, one of his favourite drinks.

The feature cocktail for the evening is the mai tai, a tribute to his movie *Blue Hawaii*. I want to have the mai tais ready for as soon as the guests arrive.

The menu for my Elvis Dinner is as follows:

Starters

Stuffed mushrooms
Shrimp cocktail
Stuffed celery

Main Course

Meatloaf with tomato sauce
Mashed sweet potatoes
Jambalaya
Collard greens with bacon
Cornbread

Dessert

Pecan pie with whipped cream
Peanut butter and banana sandwiches

Drinks

Pepsi, mai tais and oh yes, there will be wine

For the meatloaf, I mix ground beef with chopped onion, green pepper, garlic, eggs, crushed crackers and a can of tomato sauce—all according to the Mary Jenkins Langston recipe. I have made meatloaf countless times before but never with such basic ingredients. The tomato sauce has made the mixture wet, and I have never used crackers in any recipe. For anything. I feel a wave of anxiety over how this will turn out. Will my meatloaf fall apart? Will my guests hate it? Nevertheless, I shape the mixture into a loaf pan and place it in the oven. I am nervous.

Next I make the tomato sauce for the top of the meatloaf by mixing together ketchup and another can of tomato sauce. I haven't used ketchup in years. Ketchup and tomato sauce—how can this dish be good?

I try to picture Mary Jenkins Langston standing at Graceland's kitchen counter making meatloaf with the harvest gold stove in the background. I check the recipe card every thirty seconds, but I bet she could make this blindfolded.

When the timer goes, I remove the meatloaf from the oven, pour the sauce over it and bake it for an additional ten minutes.

Mary Jenkins used simple, basic ingredients for the meatloaf, and it is the same for the cornbread, collard greens and all the other dishes. All are traditional, easy-to-make home-cooked meals that Elvis enjoyed and shared around the Graceland dining table.

All of a sudden the doorbell rings. Our guests are right on time. Before I answer the door, I put on my Elvis wig, adjust my Elvis sunglasses and turn up the volume on my *Elvis Greatest Hits* to full blast. "Blue Suede Shoes" is blaring so loud they can hear it in Memphis.

When Franny and I open the front door we are greeted with deafening screams and cheers. My guests are also dressed in Elvis-inspired garb! The mai tais are poured and the dancing begins . . . *Blue blue, blue suede shoes* . . .

Dinner is a great success. The mai tais are wickedly good—dark rum, white rum, lime juice and orgeat (almond) syrup.

The meatloaf vanishes. It is flavourful, moist and juicy. My fretting over the crackers and tomato sauce was a waste of time—as is the case with most of my worrying. Elvis has been quoted as saying that he liked meatloaf more than steak because he could eat it faster. I get it.

We inhale the sweet potatoes, cornbread and jambalaya. Like Louis Armstrong's red beans and rice, this southern food is addictive—we cannot stop eating. We toast Elvis multiple times during the evening.

A big hit is the pecan pie. Just like the walnut pie I made at the Gamble House, this confection made with corn syrup and pecans is a crowd-pleaser—and the mounds of whipped cream don't hurt either.

The big surprise of the evening is the peanut butter and banana sandwiches. I make them as a joke just before the guests leave. I mix heaps of peanut butter and bananas together and spread the mixture onto white bread. I then butter the outside of the bread and fry the sandwiches just like a grilled cheese.

But as we bite into the sandwiches, we realize this is no joke. They are surprisingly good. The outside of the bread is a delightful crispy golden brown and leaves a buttery finish on the tongue. The inside of the bread is warm and moist, and the peanut butter and banana filling has a familiar sweet and savoury flavour that reaches back into my happy past. There are smiles all around. Everybody reaches for more. My good friend Dave LeBlanc remarks, "This dinner was good. But it's also the most unhealthy meal I've ever had in my life."

Elvis had an extremely close relationship with his mother, Gladys, and he was devastated when she died, in 1958. When he was growing up, Gladys used to make him peanut butter and banana sandwiches for breakfast, lunch, dinner and dessert.

Eating the food of his childhood would have been a comfort to the older, troubled Elvis.

Whereas the living room, dining room and jungle room at Graceland have a Vegas vibe, the kitchen is genuine, homely.

It is the kitchen he would want for his mother.

Graceland

3764 Elvis Presley Boulevard, Memphis, Tennessee

www.graceland.com | 800-238-2010

Information on the Elvis Honeymoon Hideaway in Palm Springs can be found at www.elvishoneymoon.com.

Dear Franny,

Visiting the Graceland kitchen was like a happy voyage back in time. Here are some '70s takeaways for us.

The visual variety in Graceland's kitchen told me that we can be individuals. We don't have to do a white-kitchen template. Let's go with something that reflects our interests and personalities.

Colour is good! Maybe not the stained-glass lampshades with apples, grapes and pears, but something that will lift our spirits. On second thought, I rather like those lampshades.

Pattern is a design element that has been passed by in contemporary kitchens but it is everywhere in the Graceland kitchen. I am not asking for a loud and brassy carpet or faux wood panelling, but pattern could be incorporated with an understated wallpaper panel or in the fabric of the kitchen chairs or the mosaic of a backsplash. Something to add cheerful visual interest.

I'm all for a television in the kitchen. I might even spend more time in this room. I find baking shows more relaxing than hockey games.

What I saw at Graceland were rooms that were over-the-top flamboyant. All of them. Except the kitchen. And when Elvis had dinner guests at Graceland, including the Beatles, he was unapologetic about serving basic southern food like fried chicken and sweet potato pie. When it came to eating and the food made in his kitchen, Elvis could be himself. He didn't have to put on an act.

Sometimes we fantasize about what we would serve Martha Stewart or Julia Child if they came to our house for dinner. Now I think we could serve them Mary Jenkins Langston's meatloaf and pecan pie. They're American classics. Elvis would approve.

You decide.

XOXO

J.

Thank ya. Thank ya very much.

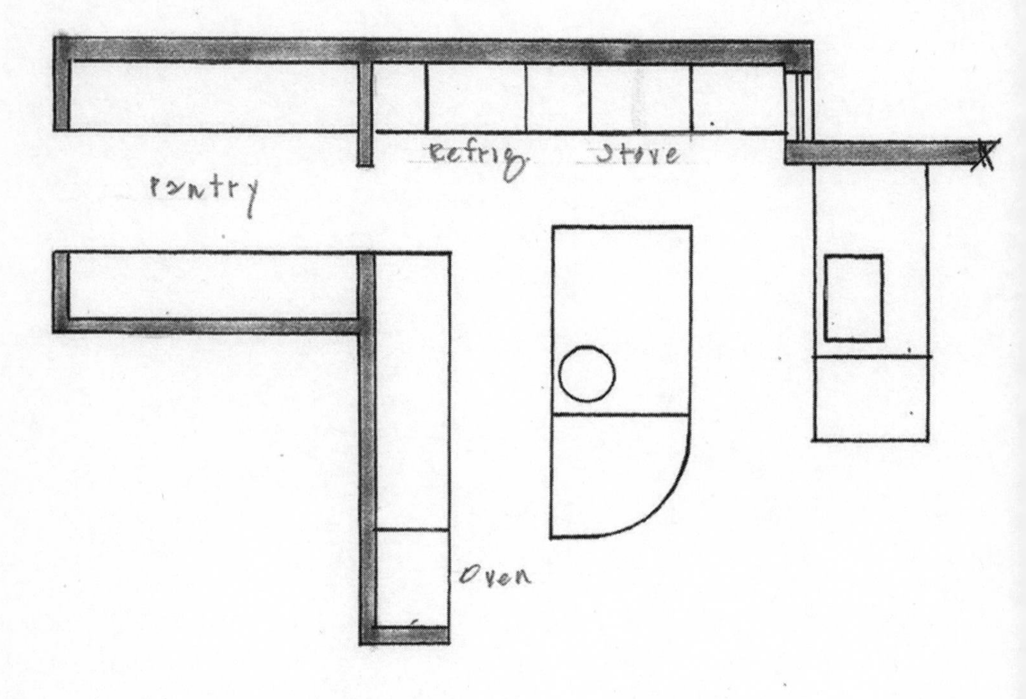
Refrig
Stove
Pantry
Oven
Plan Pearlstone Kitchen

Interior Pearlstone Kitchen

13

PEARLSTONE KITCHEN, 2016

Deep Cove, North Saanich Vancouver Island, British Columbia

My quest so far has left me with a long list of must-haves. I've seen the basic necessities of the Pilgrim kitchen at Plimoth, the innovation of the Thomas Jefferson kitchen at Monticello, the brilliant turquoise of the Lucille and Louis Armstrong kitchen and the informality of the Julia Child kitchen. They are all places where people loved to cook fabulous food. I want aspects of each in our own perfect kitchen.

But after touring far and wide and thinking hard, I believe that the number one priority for our kitchen is simplicity.

Many of my friends are turning to the West Coast to find simplicity. After slogging it out in offices all their lives, they are looking to reinvent themselves, connect with the land and find meaning on the other side of the Rockies. Ocean surf, purple mountains and marmalade skies are an irresistible lure for those searching for alternatives.

I think that the laid-back West Coast lifestyle comes from being immersed in a spectacular setting of forested mountains and ocean coastline. One cannot help but be awed by nature. The first time I visited Vancouver, my relatives walked me to the downtown beach at the shore of English Bay to watch the sunset. We were joined by what seemed like thousands. Children playing Frisbee on the beach, couples holding hands, families with arms around each other sitting against giant cedar logs—all to watch a brilliant orange sun descend into the Pacific Ocean. There is a culture there of taking time out and immersing oneself in nature.

And then there is the architecture.

Many argue that the most innovative and remarkable houses are on the West Coast. The natural landscape, agreeable climate and use of local wood and stone lend a euphoric yet casual ambience to home life unmatched in the rest of North America. Some of these houses appear as though they grow right out of the ground.

Years ago, while studying at architecture school in Vancouver, I became entranced by the magic of handmade houses. Professionals and amateurs alike built houses for themselves, some with unconventional rooflines or hanging off cliffs or up in trees. I travelled from the rugged landscape of British Columbia down to Washington State and California to search out these eccentric West Coast houses.

The handmade houses I liked best were the simple ones—a basic shed roof, recycled barnboard siding, reused windows, the components arranged not according to an architectural magazine but according to the needs and desires of the owner/builder. Modest, straightforward house construction, built by DIYers with minimal cost and complication. Maybe that's why I was so drawn to the plain architecture of the Pilgrim houses at Plimoth.

In my search for a perfect kitchen, I heard about a new house on Vancouver Island that contains a beautiful kitchen. The owners live off the land and the ocean, fire their own ceramics and craft furniture with their own hands. In the spirit of West Coast handmade houses, the owners, Joyce and Paul Pearlstone, designed the house themselves.

They also happen to be related to me and I haven't seen them for decades.

I want to know if the dream of handmade houses is still alive on the West Coast. And so I am off to Vancouver Island in search of a perfect contemporary kitchen. I am in search of simplicity.

I've always wished I could live by the ocean. Far from the tough neighbourhood in downtown Toronto where I grew up. But for some people, like Joyce and Paul Pearlstone, living by the sea is not fantasy.

Franny has joined me on this West Coast trip and I am really happy to have her in the passenger seat next to me as we search for Joyce and Paul's house. As a lawyer, Fran is the smart one between us—so she navigates and I drive. If I were navigating, we'd end up in Moscow.

We are engulfed in British Columbia's spectacular landscape of mountains and ocean as our car climbs up and down winding roads and does a balancing act along sheer drops into the ocean. We find it hard to believe that there are so many trees. There are millions, and the tree-covered mountains seem to run continuously right to the North Pole.

The route to the house takes us to the top of a cliff, where there is a small clearing surrounded by wild bushes and jagged rocks. I look around and there is no house in sight.

I park the car, and Franny and I walk along a gravel path that we assume will take us to the house. Feeling lost, we reach the edge of the cliff and look down. Below us is a path of large, flat stones that meanders down a steep slope into an enchanting garden of bamboo stalks, stone walls and flat terraces. "Oh my god," gasps Franny. "I want to live here."

I drink in the serendipitous feeling of the garden while we descend the stone steps. Japanese maples, ferns and grasses thrive among jagged rocks that have been softened by carpets of moss. A Buddha statue sits, eyes closed, hands folded, surrounded by epimedium plants. Enveloped by the peace and tranquility of the lush surroundings, this is the kind of place where I could shut myself away from the rest of the world and take up meditation.

Franny is a real gardener and she is in her glory, touching, sniffing and stroking the blooms and leaves. Farther down the hill I spot two figures digging and firming up a stone step. I get closer and I recognize my cousin Joyce and her husband, Paul, the owners of this garden paradise.

"It's so good to see you," I shout out, and there are big hugs all around.

"Johnny, we're so glad you're here," says Joyce. I love it when my relatives call me "Johnny." It makes me feel five years old again.

The last time I saw Joyce and Paul, they were bringing up two children and building careers in the scientific field. Now they have escaped the cold climate of northern Canada, retired to the glorious West Coast and built their dream house here. As I look out over the tumbling garden I can only guess that landscaping must now be their full-time occupation. But Joyce tells me otherwise.

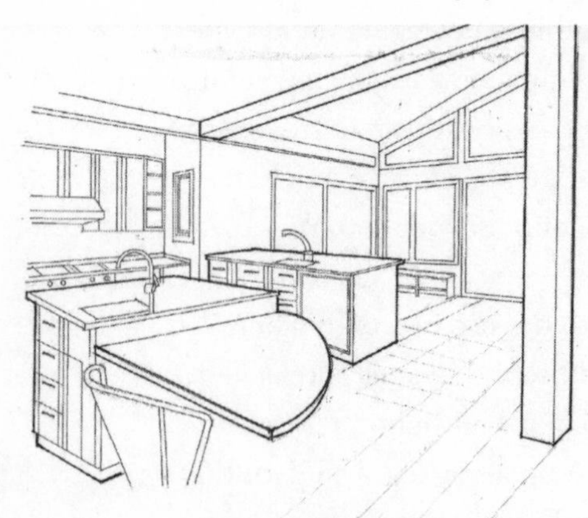

oriented to the view

"When we aren't gardening or cooking or fishing, we're busy with grandparenting duties," she says. "Otherwise, it's quite tranquil here. The only sound is the water lapping against the shore."

Along with her sister Grace (who lives close by), she is active as a board member for a chamber music society and she makes pottery. Paul spends time making furniture and wine and teaches classes in marine life at a nearby oceanic centre.

Joyce and Paul tour us through the beds of hostas and Japanese dwarf evergreens and maples, and I can tell that Franny has completely fallen in love. Joyce tells us that the garden also provides a bounty of vegetables and berries for dishes in the kitchen. But the real eye-catcher for me are the wasabi greens that grow in pots.

Wasabi! I eat this Japanese delicacy with sushi and noodles and I put it in marinades and salad dressing. It adds a jolt-up-the-nasal-passages to any dish. In Japan, wasabi grows naturally in cool mountain streambeds. Elsewhere, it's prohibitively expensive to grow the real thing, which is why the green stuff that comes with your sushi tends to be a mixture of horseradish and green food colouring. I've never before seen wasabi growing. If I can nurture this small, low-lying leafy-green plant in our kitchen garden, this alone would make our trip to the West Coast worthwhile.

After Joyce and Paul bought the land, they had to clear the site. It was up to the two of them to cut down trees, move earth and carve terraces into the slope that descends to the Pacific Ocean. A lot was done by hand.

They hired an architect initially, but weren't happy with the results. Joyce felt she had better ideas and took matters into her own hands. Even though she had no architectural training, Joyce designed the house on pieces of grid paper.

We continue to walk down the hillside and I begin to see a sloped roofline beyond the foliage. To enter the house, Franny and I step into a small vestibule that looks onto a gentle waterfall and rock garden of maidenhair ferns and spirea bushes. I'm reminded of the Kentuck Knob House and the way Frank Lloyd Wright designed low entranceways in his houses so that he could surprise visitors with the open and impressive main spaces.

Just as I am thinking this, Joyce leads me around the corner of the vestibule and BAM!

I am bowled over by a spectacular view of the Pacific Ocean and a coastal mountain range that spans the entire back of the house. In fact, it looks like there is no back to the house at all. It's as though the wall has been blown off and we are standing on a rock outcropping directly over the ocean.

I try to get my bearings. I can see that the entire main level is one open-concept room. Ahead of me is the dining and living area, and to my left is the kitchen, all overlooking the ocean and mountains.

Above me is a cedar clerestory ceiling with a line of windows that adds height to the airy interior. Sunlight streaming in from generous windows and skylights bounces off the Douglas fir posts and beams and the blond wood floors.

The posts and beams that divide up the space show off the structure that holds the house up, but they also act as picture frames enhancing the beauty of the interior. Fran remarks how the posts and beams and the big space remind her of a Haida longhouse.

The seductive views draw us deeper, into the dining and living area. Grounded by a fireplace built of slate and Douglas fir, the living area is furnished with plush couches and a baby grand piano. I admire the walls that are graced with a mixture of Haida prints and Asian art collected from past travels. Bookshelves showcase Joyce's ceramic pieces that double as both serving platters and works of art.

The main floor comprises the living spaces, while the lower level holds the bedrooms; it is cooler at night, so ideal for sleeping.

I glance down the stairs and see shoji screens, soft lighting from Japanese lanterns and an impressive collection of fossils displayed on wood bases.

Franny and I pass through the rear sliding glass doors and find ourselves on the deck overlooking the ocean. Joyce and Paul have furnished it with patio tables and chairs for al fresco dining and catching the ocean breezes. I can see that Franny is admiring the meandering paths that trail down to the seashell beach.

A boat is tied to a mooring buoy. Paul tells me that he was out just this morning trapping prawns. Other fish Paul is able to catch are salmon, ling cod, halibut and trout. To catch prawns, Paul drops a cone-shaped trap over the side of the boat to the rocky ocean floor two or three hundred feet below. The prawns, attracted by the bait, swim into the trap and are unable to escape. He tells me that the local prawn fishery is well managed, with rigorous attention to stocks and quality. "Good," I think. "Keep those little guys coming."

We're here for lunch, and Joyce interrupts my *National Geographic* reverie to tell me that it's time to start getting the food ready. I'm all for that, not only because I'm getting hungry but because I am dying to check out the kitchen.

We glide into a U-shaped ensemble of cabinets, stove, oven and central island that measures only about twelve by fifteen feet. Yet the kitchen feels much larger, because of its openness to the rest of the house and the views that seem to stretch for a thousand miles.

One of my favourite tenets of modern architectural design is the blurring of the separation between the inside and outside of a building. But the Pearlstones' house goes beyond the average house that tries to connect inside and outside with glass doors. The kitchen-dining-living area is the entire main floor. The whole space melds effortlessly with the garden and the ocean.

The stone countertops and the wood posts, bookshelves and kitchen cabinetry have an affinity with the rocks and cedar trees outside. The orientation of the structural beams, skylights and floorboards points attention to the rear of the house and the water. The back end of the house allows for 180-degree views, sounds, breezes and smells of the forest and sea. The kitchen is immersed in the ocean and mountains.

We walk over to the central island with its charcoal-coloured granite counter streaked with veins of seaweed green and sienna. This is a beautiful and functional island. Equipped with a pot sink, generous prep area and hefty butcher-block cutting board, it is a sturdy piece of furniture to facilitate Joyce's serious cooking.

Franny runs her hand over the smooth granite. She doesn't have to say a word. She is sending a mental text message. *I want one of these*.

The granite work surface cantilevers out to provide an informal eating area and additional counter space. Drawers incorporated into the island provide storage space for pots and pans.

The maple counters were designed to accommodate Joyce's height and cooking techniques. The intention was to be as ergonomic as possible and to have different heights in various working areas for kneading dough or rolling sushi.

The maple cupboard doors have a glass face to allow Joyce to display her crystal wine glasses, ceramic teapots and green teacups. The light finish and simple lines of the cupboards match the natural tones throughout the rest of the house. A roll-up door on the appliance garage conceals a space wired to operate the coffee machine, kettle and electric pencil sharpener. But like the Julia Child kitchen, not everything on the countertops is hidden from sight. Surfaces are accented with ceramic sake containers, Japanese lanterns and dramatic black serving platters.

The stainless steel refrigerator, restaurant-quality gas stove and exhaust hood are built into a wall of warm maple cabinetry.

The cupboard handles are an understated brushed chrome that matches the sheen of the appliances. The chairs are also a matching silver colour so that they almost disappear from sight. Everything is designed to enhance the view.

A tucked-away window by the stove reminds me of the miniature window beside the Pilgrim kitchen's open hearth. It not only gives the cook secret views to the ocean but also draws in fresh air and eliminates cooking odours.

Joyce has saved the most surprising feature for last: a pantry. This one is a cook's luxurious walk-in closet. The seven-by-ten-foot room of shelves is filled with jars of spices, jams, preserves, pickles, dried beans, sugars, oils and vinegars, ready to inject flavours year-round. A small fridge frees up the kitchen's refrigerator for more seasonings and sauces; this one is reserved for beverages.

As I breathe in the aromas from the spices, I think back to the pantries I have seen on my journey—the Thomas Jefferson underground addition at Monticello, the Victorian larder at Point Ellice House and the über-pantry of Georgia O'Keeffe, with her meticulously organized shelves of pickles and preserves. The need for a dedicated room for food storage declined after the 1960s. It was a combination of the popularity of processed convenience foods (no need to preserve food anymore), chest freezers and local supermarkets. In my family, we ate fresh vegetables and meats but we also consumed instant foods—we were inundated with popular-culture marketing and we went with the flow. But with the return of a desire for unprocessed, whole foods, the renewed popularity of entertaining at home and access to a greater range of ingredients from around the world, the pantry is back.

First, Joyce and I will do some prep work, chopping vegetables in the kitchen, and then we'll have starters and drinks out on the deck with Paul and Franny. Then we'll come back to the kitchen to make the main course. The menu is prawn and chicken pasta, green salad with wasabi leaves, and focaccia bread. Dessert, Joyce tells me, will be a surprise.

Joyce designates me as the vegetable prep cook for the pasta dish. Job one is chopping vegetables. She gives me specific directions: slice them on a diagonal one-eighth of an inch thick, which she says is best for stir-frying. I enthusiastically pick up a large knife and start slicing my carrots, followed by a Japanese eggplants, broccoli florets, red peppers, green peppers, onions and mushrooms.

I peek over at Joyce, who has cut chicken breasts into half-inch chunks and immersed them in a marinade of extra-virgin olive oil, minced garlic, soy sauce and mirin.

While we are cooking, we catch up on her kids, travel experiences and beloved family members long past. I ask Joyce what she did before she and Paul came to the West Coast, and she tells me that until retiring, she was a biochemist at a university doing medical research with isolated proteins and DNA. Looking around, I realize that her organizational mind is reflected in the way she has set up her kitchen.

In this kitchen, organization is key. Items are located near the area they'll be used, and surfaces are clean and uncluttered. The kitchen is designed with thirty-four drawers for easy and convenient access to utensils.

"Drawers are more expensive than cupboards," Joyce says, "but to me they're worth the extra cost. Everything is within easy reach."

When I open a couple of drawers to look for a knife, I notice that the interiors are well organized with plastic dividers—everything has an assigned place. I make a note to find dividers for the drawers in our perfect kitchen.

I am very happy when Joyce says the starter course is smoked salmon pâté. She made it with salmon Paul caught earlier in the year off Vancouver Island and smoked over local cedar. Joyce combined the smoked salmon with cream cheese, sour cream, Worcestershire sauce, dill, lemon juice and Thai chili sauce. My job is to spread the pâté on slices of cucumber (for the carb conscious) and also on toasted rounds of homemade pumpernickel rye bread.

There is much oohing and aahing as Joyce and I carry the tray of starters out to the deck. But as beautiful as the smoked salmon is to look at, it is even better to eat. Creamy, luscious and smoky, the marriage of the salmon and the accompanying flavours makes the pâté deliciously perfect. The slightly sour flavour of the rye bread and the crunch of the cucumber add to the indulgence.

Unfortunately, it seems there are fewer salmon off British Columbia this year. Just last week, Joyce tells me, the government suspended all fishing of salmon on the Fraser River. For some reason, fewer salmon are coming upriver to spawn. Knowing that there won't be any salmon fishing for a while, I try to savour the moment. I reach for another smoked salmon tidbit and say a silent prayer to the gods to bring back the salmon.

I could indulge in wine and smoked salmon forever, but Joyce rouses me and guides me back into the kitchen to prepare the main course. My eyes open wide when I see Joyce begin to dredge three dozen wild spot prawns—the same ones Paul caught that morning—in cornstarch. I love prawns, but I've never had them fresh out of the ocean.

I watch as Joyce jumps into stir-fry mode. First she sautés the marinated chicken chunks and then the prawns before removing them from the pan. Next, she turns up the heat and fries the vegetables that I have chopped on the diagonal (the sizzling sound is deafening). When the vegetables are cooked, she folds in the chicken and prawns. The aromas of the marinade, garlic and prawns drive me insane.

While Joyce is stir-frying, somehow she has also brought a large pot of water to a boil and dropped in handfuls of small twisted pasta called gemelli. I have never seen these spiral-shaped darlings and I am anxious to try them.

I help Joyce set the indoor table with her mingei ceramic platters. We lay out a dazzling display of prawn and chicken pasta, summer green salad and homemade focaccia.

I waste no time digging in. I inhale the aromas of garlic and the sea as I enjoy the tender chicken chunks that are spiked with soy and mirin. The twirly gemelli are perfectly al dente. Gemelli, you are my new best friend. The fresh vegetables have a nice crunch and have been blessed with a touch of grated Parmesan. I especially enjoy the homegrown kale that infuses the dish with a taste of the garden.

Sweet and succulent, with a delicate texture, B.C. spot prawns offer an intense right-out-of-the-ocean flavour.

But the stars of the show are the wild B.C. spot prawns. Sweet and succulent, with a delicate texture, they offer an intense right-out-of-the-ocean flavour. The pink-and-white crescents marry beautifully with the chicken and a hint of garlic. I decide that there is nothing better in the world than a forkful of prawn, kale and gemelli brushed with Parmesan and followed with a sip of chilled white wine.

The salad dressing of Dijon, olive oil, balsamic and garlic coats the summer salad of greens, wasabi leaves and tomatoes from the garden. From my experience at the Georgia O'Keeffe luncheon, I am reminded that freshly harvested vegetables taste so much better than produce from the store that has been sitting for days after making a long journey from the fields.

As we eat, there is much laughing and conversation, and many compliments to the chef. Together, the spectacular surroundings, the kitchen and the food have induced an aura of euphoria.

Over my feigned protests ("Oh no, I couldn't possibly"), Joyce brings out dessert. Homemade chocolate banana cake, served with espresso ice cream and homegrown strawberries. It's thick, decadent and full of flavour. This is a perfect meal.

I glance over at Franny, and she looks at me with great affection. Or does she? I detect that what she is actually doing is discreetly checking to see if anything is stuck in my teeth—one of her favourite pastimes when we are out for dinner.

"You're good," she says.

As we sit around the table enjoying our dessert, we compare the pros and cons of living in the Pacific Northwest. Let's see. Pros: the ocean . . . the mountains . . . and the seafood . . . and the trees . . . and the sunsets . . .

Cons: zero.

Dear Franny,

Thanks for coming out to visit the Pearlstone kitchen with me. It was wonderful for us to see members of my family, and I could also see that you were totally into their garden, house and kitchen. Here are some thoughts for a perfect contemporary kitchen.

The kitchen as the great room of the house. One big room that spreads into the dining and living area and into the garden and across the deck and into the ocean. It is all one space—inside and outside.

A window over the sink is a must for any kitchen. We don't have a view of the Pacific Ocean, but a window is always nice to enjoy a view, watch kids playing or do some birdwatching while washing the dishes.

Walking into a big pantry makes me feel good about cooking. A+ for creative cooking. This is like having a big walk-in closet but with our food.

The Pearlstone kitchen, like the Julia Child kitchen and the Kentuck Knob kitchen, is organized according to the needs of the cook.

I could see that lots of drawers could make cooking more efficient. It makes sense to avoid reaching into the back of the lower cupboards. Anything to save our backs.

I liked the appliance garage a lot. Hides small appliances, keeps them plugged in and avoids having to lug them in and out. Good place to charge the cell phone too.

The silver furniture blends in with the stainless steel appliances. Visually it makes the furniture disappear and gives an illusion of more space.

We don't live by the ocean and we don't have a spectacular view. But we can still have a calming contemporary kitchen with function, simplicity and a happy spirit.

Oh, and those wasabi greens? We gotta have those!

XOXO

J.

EPILOGUE

Dear John,

Thank you for this book. It is amazing. There are so many ideas for us to use and adapt from all of these famous kitchens.

But, John. I am not sure we can build a "perfect" kitchen. I know we can do a great kitchen and one we'll be very happy with for years. Even those super-kitchens you designed for clients with unlimited budgets had, in the end, some pros and cons.

You've told me many times that the best architecture involves designs that work with the existing constrained conditions. In our case, we must work within our ten-by-twelve space, the placement of our kitchen at the front of the house and the wide window that overlooks the street.

We don't have to jump in and build a spanking new shiny kitchen. We can start with a few ideas and let things develop organically. We can see what works and what doesn't and keep building on that. You wrote that the Childs set up nine kitchens before settling on the design in Cambridge.

I'm really happy that your creativity button is all lit up from cooking so many different dishes on this journey. I can hardly wait for some sensational dinners! Where do we get those Jefferson white eggplants? How about that wasabi salad?

I'm anxious to try out the organizational ideas. I like how Julia Child made cooking easier for herself by installing a magnetic knife holder at the ready, and how her favourite pots and pans each had their designated places on her pegboard. Mrs. Hagan at Kentuck Knob had everything within a two-foot reach.

And I was fascinated with the different ways Jefferson, the O'Reillys, the Gambles and the Pearlstones designed their kitchens to make beautiful dinners for their friends and family. We'll do the same—dinners, picnics, birthdays, weddings—and you know I love hors d'oeuvre parties!

I know we're going to love our kitchen.

XOXO

Franny

POSTSCRIPT

THANK YOU FOR READING this book. I had so much fun writing it. I hope it sheds some light on your house, your kitchen and your food.

Kitchens, like people, come in all shapes and sizes. Some people like cluttered kitchens, others like clean, austere surfaces, some like closed off and square, some like round and open to the living space. There is not one template for a perfect kitchen. Whatever works for you—whether it is for cooking, eating, entertaining or as a fashion statement—I hope that you enjoy your kitchen.

What does the future look like for the North American kitchen?

Kitchen islands will morph into the focal points not only of the kitchen but of the entire ground floor of the house. In the same way Lucille Armstrong made a design statement with her curvy turquoise cabinets, the island will provide glitz, with under-counter appliances, televisions, computers, compact cultivators to grow herbs and salad greens and, of course, be the family dining table. It will become more beautiful and spread right into the living area.

From faucets to fridge to coffee maker, technology has entered the house like a tsunami, and the kitchen will be the nerve centre.

Built with interior cameras and LED screens, the refrigerator will track what's inside, how much and how fresh. No more running out of milk. When you are running low, the smart fridge will contact the grocery store to make a delivery.

Most of all, the kitchen will continue to be the most important room in the house—and more. As life becomes faster and family members become increasingly isolated in their private rooms, the space where we come together for a shared meal will become more treasured. Other separate rooms will become obsolete—the dining room, living room, family room—the walls will disappear. The kitchen will take over the social spaces of the house. We can already see it happening in the Pearlstone House, where the kitchen has become the great room combining the social and food-preparation activities of the house. The kitchen expands inside and to the exterior. We have come full circle back to the Pilgrim kitchen at Plimoth.

It may not seem rational for a late-middle-aged man to have been criss-crossing North America on his own to look at kitchens. I met wonderful people on this journey and I learned a lot about historic houses, kitchens and food. I also learned why I like to visit kitchens so much.

Do you ever wonder why at a party most people gather in the kitchen?

My answer: There is less anxiety there.

Sure, it's where the food is, but there's more to it than that. At a party where you don't know anybody, it's much easier to start a conversation in the kitchen. Guests don't have to sit in the oh-so-important living room and impress each other with clever conversation and job titles.

The kitchen is informal, friendly, and it doesn't matter if you spill your red wine—the kitchen is where you'll find the paper towels. There are always little jobs to do in the kitchen to help the host or hostess, which also leads to friendly chatter. You can move around, jump in and out of conversations, talk about what really matters—the food. At a party, it is the kitchen where we go to relax, laugh and eat. It is a great anxiety-reducer and social equalizer.

Maybe, in the end, what we are seeking in the kitchen is peace of mind, comfort, happiness.

And memories.

I am looking for my grandmother to make me cinnamon toast and hot chocolate for breakfast on a dark and cold morning before school.

I am looking to stay up with my mom till three in the morning, rolling sushi on New Year's Eve.

I am looking for my parents, aunts and uncles so that we can eat, drink and laugh together.

God, I miss them. I truly believe that on the day I die, I will see them again. And I know that when I find them, it'll be in the kitchen.

J.

RECIPES

THE FOLLOWING ARE GIVEN in their original formats. I hope you agree that the different styles and formats give a small insight into the personality of the author, the type of kitchen where they cooked and their historical period. Together, they reflect the evolution of recipe writing, kitchens and cooking in North America. Further recipes relating to this book can be found at www.johnotahome.com.

Chapter 1: PILGRIM KITCHEN

To boyle a Ducke with Turneps.

Take her first, and put her into a potte with stewed broth, then take perselye, and sweete hearbs, and chop them, and perboyle the roots very well in an other pot, then put unto them sweet butter, Cynamon, Gynger, grosse Pepper and whole Mace, and so season it with salt, and serve it upon soppes.

From *The Good Huswifes Jewell* by Thomas Dawson, 1596.

To roste a Quaile.

Let his legs be broken, and knot one within another, and so roste him.

From *The Good Huswifes Handemaide for the Kitchin*, 1594.

Sauce for a Quaile, Raile, or big bird.

Sauce for a Quaile, Raile, or any fat big bird, is Claret Wine and Salt mixt together with the gravie of the Bird; and a few fine bread-crumnes well boild together, and either a Sage-leafe, or Bay-leafe crusht among it according to mens tasts.

From *The English Huswife* by Gervase Markham, 1615.

Chapter 2: MONTICELLO KITCHEN

TO DRESS SALAD.

To have this delicate dish in perfection, the lettuce, pepper grass, chervil, cress, &c. should be gathered early in the morning, nicely picked, washed, and laid in cold water, which will be improved by adding ice; just before dinner is ready to be served, drain the water from your salad, cut it into a bowl, giving the proper proportions of each plant; prepare the following mixture to pour over it: boil two fresh eggs ten minutes, put them in water to cool, then take the yolks in a soup plate, pour on them a table spoonful of cold water, rub them with a wooden spoon until they are perfectly dissolved; then add two table spoonsful of oil; when well mixed, put in a teaspoonful of salt, one of powdered sugar, and one of made mustard; when all these are united and quite smooth, stir in two table spoonfuls of common, and two of tarragon vinegar; put it over the salad, and garnish the top with the whites of the eggs cut into rings, and lay around the edge of the bowl young scallions, they being the most delicate of the onion tribe.

From *The Virginia House-wife* by Mary Randolph (Washington, DC, 1824).

Chapter 3: HERMANN-GRIMA HOUSE KITCHEN

OKRA SOUP.

Cut up in fine slices two soup plates of okra, and put into a soup pot with 5 quarts of water and a little salt, at 10 o'clock; at 11 o'clock put your meat into the soup pot; at 12 o'clock peel a soup plate and a half of tomatoes, and after straining them through a cullender throw them into the soup pot; then season with pepper and salt. Allow all the ingredients to boil till 3 o'clock, when it is fit to be served up.

Chapter 4: POINT ELLICE HOUSE KITCHEN

Hindoostanee Curry.

Time, nearly two hours.

Two pounds of meat (beef, veal, or any other you prefer); one pint and a half of water; four tablespoonsful of curry powder; two onions; one root of garlic; three ounces of butter; two tablespoonsful of cocoa-nut milk—or good cream; ten almonds; six cloves; a blade of mace; a small piece of cinnamon, and a few cardamom seeds; juice of one lemon.

Boil the meat in a pint and a half of water, till about half done, then take it out, and skim the broth; and put to it the cloves, mace, cinnamon, and cardamom seeds. Cut the meat into small square pieces, roll them in the curry powder, and fry them a nice brown in some butter. Cut up the onions and root the garlic, and fry them also until brown, but separate from the meat. Then add the whole to the broth with the cocoa-nut milk or a little good cream, and the almonds blanched and pounded. Cover the pan closely over, and let it stew gently over a slow fire until all are well mixed and very hot; and just before serving squeeze in the juice of a lemon.

From *Warne's Model Cookery and Housekeeping Book* by Mary Jewry (London and New York, 1868). This cookbook was used by Caroline O'Reilly and can be found in the Point Ellice House archives.

Chapter 5: TENEMENT KITCHEN

Matzo Ball Soup

Matzo Balls

4 eggs
2 tablespoons schmaltz or vegetable oil
¼ cup club soda or chicken broth
1 cup matzo meal
Salt and pepper

Mix eggs well with a fork. Add Schmaltz or oil, soda water or chicken broth, matzo meal, salt and pepper and mix. Refrigerate for several hours.

Wet hands in cold water and make a dozen balls the size of Swedish meat balls.

Bring water to a boil in a large pot. Add salt and place matzo balls into the water. Cover and simmer for about thirty minutes.

Chicken Soup

4 pound chicken
4 quarts of cold water
6 carrots, 6 celery stalks, 1 onion chopped in rough chunks
Salt, pepper
Handful of fresh dill

Place chicken in large pot and add water just to cover carcass. Add vegetables chunks and dill.

Bring to a boil. Lower the heat. Bring to a slow simmer for 2½ hours.

Skim chicken fat as it forms at the surface.

After 2½ hours, strain the soup and separate stock from the bones.

Strip meat from carcass and place in refrigerator. Discard bones and cooked vegetables.

Ladle hot, clear soup into soup bowl.

Add three matzo balls.

Garnish with carrot slices and sprig of dill.

Serve.

By kind permission of Anne Lewison, based on her grandmother's recipe.

Chapter 6: GAMBLE HOUSE KITCHEN

AUTOMOBILE SALAD. Miss W. I. Puls, 824 Tenth Street, Riverside, Cal.—Cut into small pieces four medium-sized tomatoes, draining off the juice and rejecting it from the salad; two medium-sized heads of lettuce, four stalks of celery and one-half cup pickled olives. Mix thoroughly and put together with the following dressing: Beat one egg until creamy; pour over it four tablespoons vinegar, scalding hot, stirring constantly. Place dish in hot water over fire and stir constantly until mixture thickens. Remove from fire and add one teaspoon butter and stir until melted. Add one-half teaspoon mustard, one-half teaspoon salt, one-quarter teaspoon pepper and dilute with enough sweet cream to make the dressing the consistency of cream. Add two tablespoons Underwood deviled ham. Garnish with red nasturtium blossoms.

From *Los Angeles Times Cook Book—No. 2* (Los Angeles: Times-Mirror, 1905).

Chapter 7: SPADINA HOUSE KITCHEN

SCOTCH SHORTBREAD
(No. 3)

¼ pound fresh butter
2 ounces fine sugar
½ ounce cornstarch
6 ounces *Five Roses* flour

Knead cornstarch and sugar into the butter, then gradually knead in flour. Roll out into a round. Pinch the edges with fore-finger and thumb, prick over top with fork, cut in eight. Place on baking dish and bake in moderate oven 20 minutes. Leave on tin to harden.

Adapted from *Five Roses Cook Book* (Winnipeg and Montreal: Lake of the Woods Milling, 1915).

Chapter 8: GEORGIA O'KEEFFE KITCHEN

Corn Soup

SERVES 3

2 cups corn kernels
2 cups milk
1T. minced onion
1T. soup mix (optional)
Herb salt, to taste
Chives or parsley, as garnish

1. Put the fresh, raw corn in a blender container. Add the milk, onion, soup mix, and herb salt. Blend at the highest setting for about 15 seconds.

2. Use a pestle to push the liquid through a sieve into a pan. Heat the soup slowly, stirring it continuously as it thickens.

3. Serve immediately, do not simmer. Garnish with finely chopped chives or parsley.

Soup Mix

1½ cups powdered milk
1 cup soy flour
½ cup kelp
1 cup brewer's yeast

Measure these ingredients into a glass storage container. Cover with a tight-fitting lid and shake the jar to combine the ingredients thoroughly. Add this to creamed soups and breads to taste.

Adapted by Santa Fe School of Cooking from *A Painter's Kitchen: Recipes from the Kitchen of Georgia O'Keeffe* by Margaret Wood (Santa Fe: Museum of New Mexico Press, 2009).

Chapter 9: FRANK LLOYD WRIGHT KITCHEN

Baked Alaska

This recipe appears in my friend Sylvia Lovegren's highly recommended book *Fashionable Food: Seven Decades of Food Fads*.

She writes: "Said *641 Tested Recipes from the Sealtest Kitchens* (1954), 'Elegant, glamourous Baked Alaska is simple as ABC if you'll rigidly follow these three musts: a. the ice cream must be very hard; b. the ice cream must be sealed in with a fluffy egg meringue; c. the oven must be piping hot. The result—out of this world eating!' You may increase the egg whites to five and the sugar to eight tablespoons and pipe pretty designs over the top of the meringue just before baking. But only attempt this if you are expert—this is no time to experiment while the ice cream melts."

4 egg whites
6 tablespoons sugar
¼ teaspoon salt
½ teaspoon vanilla extract
1 (9-inch) layer sponge cake or genoise, about 1 inch thick
½ cup currant jelly
1 quart hard strawberry ice cream
2 large eggshell halves, with neat edges, washed and dried
3 tablespoons brandy, heated just before using

Preheat the oven to 450 deg F. Beat the egg whites until stiff but not dry. Beat in the sugar and salt, adding them gradually but quickly, until the meringue is thick and glossy. Beat in the vanilla.

Cover a bread board with baking parchment. Place the cake

on the board and spread the jelly on the cake. Mound the ice cream on the cake, keeping ½ to 1-inch edge clear on the cake. Frost with the meringue, working quickly. Be sure every bit of the ice cream is covered for best insulation. Nestle the eggshell halves open side up on top of the meringue. Pop the cake into the oven until the meringue has just browned slightly, 3 to 5 minutes. Quickly slip the Baked Alaska onto a chilled platter. Fill the eggshells with the heated brandy, set alight, and rush the Baked Alaska to the table.

Note: For tender Fifties sophisticates who didn't use likker, sugar cubes soaked in lemon extract replaced the egg shells and brandy. Makes 10 to 12 servings.

(John Ota notes: In making my baked Alaska, I put it under the broiler to brown—not in the oven at 450 deg. I did not use currant jelly. I did not use egg shells but poured flaming rum over the meringue.)

Adapted from *Fashionable Food: Seven Decades of Food Fads* by Sylvia Lovegren (University of Chicago Press, 2005). Used by permission of the author.

Chapter 10: JULIA CHILD KITCHEN

Cheese Soufflé

FOR A 1-QUART BAKING DISH 8 INCHES ACROSS, SERVING 4

2 Tbs finely grated Parmesan or other hard cheese
2½ Tbs butter
3 Tbs all-purpose flour
1 cup hot milk
Seasonings: ½ tsp paprika, speck of nutmeg, ½ tsp salt, and
3 grinds of white pepper
4 egg yolks
5 egg whites (⅔ cup)
1 cup (3½ ounces) coarsely grated Swiss cheese

Special Equipment Suggested:

A buttered baking dish 7½ to 8 inches top diameter, 3 inches deep; aluminum foil [24 inches long, folded in half lengthwise]; a small saucepan for the milk; a heavy-bottomed 2½-quart saucepan; a wire whisk, wooden spoon, and large rubber spatula; egg-white beating equipment

Preliminaries. [Butter a 3-inch strip along the folded edge of the foil. Wrap the foil collar around the soufflé dish; secure the collar by inserting a straight pin head down—for easy removal. Cut a length of kitchen twine about 2½ to 3 feet and tie it tightly around the top of the soufflé dish to hold the foil firmly in place. Carefully sprinkle 2 Tbs of the Parmesan cheese inside the dish, covering the sides and the bottom and also covering the buttered portion of the foil.] Preheat oven to 400°F, and set the rack in the lower third level. Measure out all the ingredients listed.

The white sauce—béchamel. [Heat the milk in the small saucepan over moderate heat. Do not allow it to boil. Melt the butter in the 2½-quart saucepan, then blend in the flour with a wooden spoon to make a smooth, somewhat loose paste.] Stir and cook the butter and flour together in the saucepan over moderate heat for 2 minutes, without coloring. Remove from heat, let cool a moment, then pour in all the hot milk and whisk vigorously to blend. Return to heat, stirring with a wooden spoon, and boil slowly 3 minutes. The sauce will be very thick. Whisk in the seasonings, and remove from heat.

Finishing the sauce base. One by one, whisk the egg yolks into the hot sauce.

The egg whites. In a clean separate bowl with clean beaters, beat the egg whites to stiff shining peaks.

Finishing the soufflé mixture. Scoop a quarter of the egg whites on top of the sauce and stir them in with a wooden spoon. Turn the rest of the egg whites on top; rapidly and delicately fold them in, alternating scoops of the spatula with sprinkles of the coarsely grated cheese—adding the cheese now makes for a light soufflé.

**Ahead-of-time note:* You may complete the soufflé to this point ½ hour or so in advance; cover loosely with a sheet of foil and set away from drafts.

Baking—25 to 30 minutes at 400°F and 375°F. Set in the preheated oven, turn the thermostat down to 375°F, and bake until the soufflé has puffed 2 to 3 inches over the rim of the baking dish into the collar, and the top has browned nicely.

Serving. As soon as it is done, remove the collar, then bring the soufflé to the table. To keep the puff standing, hold your serving spoon and fork upright and back to back; plunge them into the crust and tear it apart.

Chapter 11: LOUIS ARMSTRONG KITCHEN

POP'S FAVORITE DISH

By - Louis and Lucille Armstrong

CREOLE RED BEANS (KIDNEY) AND RICE

(Use 2 qt. pot with cover)
1 lb. Kidney Beans
1/2 lb. Salt Pork (Strip of lean, strip of fat) (Slab Bacon may be used if preferred)
1 small can of tomato sauce (if desired)
6 small Ham Hocks or one smoked Pork Butt
2 onions diced
1/4 green (bell) pepper
5 tiny or 2 medium dried peppers
1 clove garlic - chopped
Salt to taste.

PREPARATION

Wash beans thoroughly, then soak over night in cold water. Be sure to cover beans. To cook, pour water off beans, add fresh water to cover. Add salt pork or bacon, let come to a boil over full flame in covered pot. Turn flame down to slightly higher than low and let cook one and one-half hours. Add diced onions, bell pepper, garlic, dried peppers and salt. Cook three hours. Add tomato sauce, cook one and one-half hours more, adding water whenever necessary. Beans and meat should always be just covered with water (juice), never dry. This serves 6 or more persons.

To prepare with Ham Hocks or Pork Butts. . . Wash meat, add water to cover and let come to a boil in covered pot over medium flame. Cook one and one-half hours. Then add beans (pour water off), add rest of ingredients to meat. Cook four and one-half hours. Add water when necessary.

SUGGESTIONS

For non pork eaters, chicken fat may be used instead of salt pork. Corned beef or beef tongue may be used instead of ham hocks or butts.

RICE

2 cups white rice
2 cups water
One teaspoon of salt
One pot with cover

Wash rice thoroughly, have water and salt come to a boil. Add rice to boiling water. Cook until rice swells and water is almost evaporated. Cover and turn flame down low. Cook until rice is grainy. To insure grainy rice, always use one and one-half cups water to one cup of rice.

TO SERVE

On dinner plate—Rice then beans, either over rice or beside rice as preferred. . . Twenty minutes later—Bisma Rex and Swiss Kriss.

Chapter 12: ELVIS PRESLEY GRACELAND KITCHEN

Meat Loaf

Mix:

2 lbs. ground beef
1 cup onion
1 bell pepper
2 cloves of garlic
3 eggs
1 pkg (4 oz) crackers crushed
1 8 oz. can tomato sauce

Bake in 350 degree oven.

Sauce:

2—8 oz. cans tomato sauce
½ cup ketchup

Pour over meatloaf when nearly done and return to oven.

Many thanks to Graceland for this authentic recipe from Elvis's cook, Mary Jenkins Langston.

Chapter 13: PEARLSTONE KITCHEN

Smoked Salmon Pâté

Ingredients:

- 1 brick cream cheese (softened at room temperature)
- 2 tablespoons sour cream
- 2 cups smoked salmon (Deep Cove)
- Dash lemon juice
- Tabasco or Thai chili sauce to taste
- 2 tablespoons finely chopped fresh dill
- Dashes of Worcestershire sauce

Preparation:

1. Place all the ingredients in a bowl.
2. Stir together.
3. Taste for seasoning.
4. Transfer to a bowl and keep refrigerated until ready to serve.
5. Serve on toasted rounds of rye bread and cucumber slices.

By kind permission of Joyce Pearlstone.

ACKNOWLEDGEMENTS

I DIDN'T KNOW that writing a book was a team sport.

But it is.

When my friend Jim Polk heard I was writing a manuscript, he said, "You'll have to have a good team." I didn't know what he was talking about. I thought writing was a solitary activity. I was so wrong. I have been lucky to be part of a great team.

There are so many people to thank for their help with this book. Unfortunately, there is not enough space to name them all. But I hope you know who you are.

Number one, I thank Frances Rowe, who is my wife, partner in life, social convenor, CFO and psychiatrist. For the last few years, this person has been living with a maniac, demanding complete silence in the house and disrupting planned vacations at the very last second as I search for another kitchen.

I am most grateful to Robert McCullough, publisher of Appetite by Random House, Penguin Random House Canada, for taking a chance with an unknown author. My literary agent, Carolyn Swayze, and associate agent Kris Rothstein encouraged me from the start, gave me sage advice and have always been in my corner. Cynthia

Good and her class at Humber College connected me with Maria Coletta McLean. Maria taught me invaluable lessons on how to write and what to read and introduced me to Carolyn Swayze. Shaun Oakey, vigilant copy editor and guardian angel, caught and corrected countless mistakes and made valuable suggestions for improving the text. Thank you to PRHC designer, the brilliant Leah Springate, for beautifying this book, inside and out. Thanks to Ward Hawkes, Lindsay Vermeulen and all the team at PRHC for their design and publishing support. For their relentless positive advice, I am grateful to friends Jim Polk, Eddy Yanofsky, Anne Lewison, David Lillico, Tim Lindsay, Fred Cane, Dale Hauser and Jane Simmons. Paul McLaughlin, Lyn Hamilton and my first grade teacher, Mrs. Lennox, have been my life writing mentors.

Thank you to Paula Johnson, Margaret Wood, Emily Butler, the Palumbo Family, Karen Edwards, Miriam Bader, Becky Nicolaides, Sheryl Scott, Charles Perry, Kathleen Wall, Eleanor Gould, Jennifer Dyer, Gail Simpson, Nidhi Sheth, Ed Lyons, Joyce and Paul Pearlstone, Grace Kamitakahara, Katherine Redd and the late David Reese, who are among the fabulous cooks, historians, curators, tour guides, writers and architects who devote their lives to preserving historic houses and promoting culinary history. Sylvia Lovegren and members of the Culinary Historians of Canada were a constant source of information and encouragement.

My brother Chris called me from LA continuously with creative ideas on writing and marketing, and accompanied me on cooking trips.

On every team there is a Most Valuable Player and here that person is my editor, Tim Rostron. Tim led a first-time author through the labyrinth to transform the manuscript into a book. I thank him for his patience, ideas, vision and his ability to make this all happen—*MVP, MVP, MVP* . . .

And finally, and most deeply, I thank my friends, neighbours, relatives and acquaintances who asked about the book, gave me encouragement and shared their thoughts with me about the kitchen.

Thank you. We are all members of Team Kitchen.

J.

SOURCES AND FURTHER READING

General

Excellent books on the history of the house are *The Food Axis: Cooking, Eating, and the Architecture of American Houses* by Elizabeth Collins Cromley (Charlottesville and London: University of Virginia Press, 2010) and Gwendolyn Wright's *Building the Dream: A Social History of Housing in America* (New York: Pantheon Books, 1981). A good guide to understanding the materials and structure of houses is *The Vintage House: A Guide to Successful Renovations and Additions* by Mark Alan Hewitt and Gordon Bock (New York and London: W.W. Norton, 2011).

Additional articles and recipes can be found at www.johnotahome.com.

Chapter 1: PILGRIM KITCHEN

The Plimoth "Hardcore Hearth Cooking" class, including recipes, cooking techniques and history, was led by Kathleen Wall, colonial foodways culinarian at Plimoth Plantation. Information on the early life of the Puritans can be found in *Plimoth Plantation: A Story of Two Cultures* by Linda Coombs et al. (Lawrenceburg, IN: R. L. Ruehrwein, 2008), distributed by Plimoth Plantation. Details on house construction, food and cultivating maize come from educational essays on the Plimoth Plantation website, www.plimoth.org. A comprehensive overview of Puritan houses can be found in

Revelations of New England Architecture by Jill Grossman, with photographs by Curt Bruce (New York: Grossman, 1975.) Pilgrim food is described at "The History of Thanksgiving," Colorado Women's College, University of Denver, https://mysite.du.edu/~fdallas/its2410/hw5_thanksgiving/index.html. Information on growing and eating pumpkins can be found at the Colonial Williamsburg website, www.history.org.

Chapter 2: **MONTICELLO KITCHEN**

The cooking class "Jefferson's Table: Recipes from the Monticello Kitchen" was led by Eleanor Gould, curator of gardens at Monticello. An excellent source for recipes and information is *Dining at Monticello: In Good Taste and Abundance*, edited by Damon Lee Fowler (Chapel Hill: University of North Carolina Press for Thomas Jefferson Foundation, 2005). The Monticello kitchen is described at "French Cuisine in a Virginia Kitchen" on the Monticello website, www.monticello.org. Historical recipes are from *The Virginia Housewife* by Mary Randolph (Washington, D.C., 1824), https://catalog.hathitrust.org/Record/009728491. Architectural background on Monticello can be found in *Jefferson's Monticello* by William Howard Adams (New York: Abbeville, 1983) and *Pride of Place: Building the American Dream* by Robert A. M. Stern (Boston: Houghton Mifflin, 1986) as well as *Thomas Jefferson* (*Time* special issue, 2015), Thomas Jefferson America's Enduring Revolutionary, © 2015 Time Inc. Books, an imprint of Time Inc Books, 1271 Avenue of the Americas, 6th floor, New York, NY 10020.

Further information on Monticello can be found at its website, www.monticello.org.

Chapter 3: **HERMANN-GRIMA HOUSE KITCHEN**

Information on the kitchen, as well as cooking demonstration and recipes, was provided by Jennifer Dyer, education officer at the Hermann-Grima House. Jennifer Dyer is no longer at the house and, since my visit, there have been changes in staff and management. Additional details came from chief curator Katie Burlison. Further information on the Hermann-Grima House can be found at its website, www.hgghh.org. More about the history of the house and New Orleans can be found in *Luxury, Inequity, and Yellow Fever* by Kerri McCaffety (New Orleans: Vissi d'Arte Books, 2014). Information on the history of the New Orleans market is at www.frenchmarket.org and in "French Market Celebrates 200th Anniversary" by John Magill, *Preservation in Print*, vol. 18, no. 4 (May 1991). A brief history of Cajun and Creole food can be found at http://destinationsdiva.blogspot.ca/2017/09/a-brief-history-of-creole-and-cajun.html. Historic recipes are found in *Creole Cookery* by the Christian Woman's Exchange (1885; reprint, Gretna, LA: Pelican, 2005). Hearth cooking descriptions are in the book *Hearth and Home: Women and the Art of Open-Hearth Cooking* by Fiona Lucas (Toronto: James Lorimer, 2006).

Chapter 4: POINT ELLICE HOUSE KITCHEN

The history of the house and kitchen was presented by Gail Simpson, president of the Point Ellice House Preservation Society, and the Indian cooking demonstration was given by chef Nidhi Sheth. Further information on the house can be found in *Clue to a Culture: Food Preparation of the O'Reilly Family* by Virginia A. S. Careless (Victoria, BC: Royal British Columbia Museum, 1993) and at www.pointellicehouse.com. Background on Chinese immigrants is taken from "The Chinese Experience in 19th Century America," University of Illinois at Urbana-Champaign.

Chapter 5: LEVINE TENEMENT KITCHEN

Miriam Bader, director of education at the Tenement Museum, described to me the history and way of life in the turn-of-the-nineteenth-century tenements. Anne Lewison gave me an unforgettable cooking lesson. Information on the Levine family and life in a sweatshop can be found at www.tenement.org. A beautifully written and informative book on tenement life and food is *97 Orchard: An Edible History of Five Immigrant Families in One New York Tenement* by Jane Ziegelman (New York: HarperCollins, 2010); another source is *Urban Appetites: Food and Culture in Nineteenth-Century New York* by Cindy R. Lobel (University of Chicago Press, 2014). A thorough architectural history of the building is *Biography of a Tenement House in New York City* by Andrew S. Dolkart © 2012 Andrew S. Dolkart and the Center for American Places at Columbia College Chicago. Distributed by the University of Virginia Press.

I consulted the articles "Chicken Soup with Matzah Balls" by George Erdosh in *Jewish Magazine* (January 2001), www.jewishmag.com/39mag/soup/soup.htm, and "Jewish Chicken Soup with Matzo Balls . . . the Real Jewish Penicillin" at www.girlandthekitchen.com/jewish-chicken-soup-with-matzo-balls-the-real-jewish-penicillin/.

Chapter 6: GAMBLE HOUSE KITCHEN

Details on the history and building of the Gamble House can be found at https://gamblehouse.org; in *The Gamble House: Building Paradise in California* by Edward R. Bosley, Anne E. Mallek, Ann Scheid and Robert Winter (The Gamble House/University of Southern California School of Architecture and CityFiles Press, 2015); and in *The Gamble House* by Linda G. Arntzenius (Los Angeles: University of Southern California School of Architecture, 2000). Further sources are the chapter "A Leaven in the Blood: Greene and Greene and the California Bungalow" in *Architecture of the Sun: Los Angeles Modernism, 1900–1970* by Thomas S. Hines (New York: Rizzoli, 2010). Background on the family can be found in an entertaining article, "The Gambles and The Smiths" by Anne Mallek, *Update* (Spring 2015). Information on Arts and Crafts kitchens was drawn from *Bungalow Kitchens* by Jane Powell and Linda Svendsen (Layton, UT: Gibbs Smith, 2000) and the article "Coming Up for Air: The Urban Ontario Kitchen Emerges from the Victorian Basement" by Paul Sharkey, *Culinary Chronicles* (Spring 2007). Information on 1908 cuisine was sourced from correspondence and interviews with Charles Perry, president of the Culinary Historians of Southern California; from Perry's article "The Salad Eaters," *Los Angeles Times*, April 23, 2003; and the *Los Angeles Times Cookbook—No. 2* (Los Angeles: Times-Mirror, 1905). Information on the history of Pasadena and Los Angeles suburbs can be found in Becky Nicolaides's 1989 paper "City on the Last Hill," unpublished paper in author's possession. Background on Pasadena picnics is found in *Homesickness: An American History* by Susan J. Matt (New York: Oxford University Press, 2011). Historic photographs of early twentieth-century picnics can be found at the Library and Archives, Pasadena Museum of History.

Chapter 7: SPADINA HOUSE KITCHEN

Karen Edwards, Doug Fyfe, Karen Black and Fiona Lucas shared their knowledge, information and documentation of the house, and Ed Lyons, my historic baking mentor, taught me to make early twentieth-century shortbread. The history of Spadina is outlined in the book *Spadina Museum: Historic House & Gardens* (City of Toronto, 2008), distributed by Spadina Museum. Information on the icebox is from the essay "Spadina's Magnificent Eureka Refrigerator" by Fiona Lucas, unpublished paper in author's possession. Information on the history of the kitchen can be found in Paul Sharkey's article "Coming Up for Air: The Urban Ontario Kitchen Emerges from the Victorian Basement," *Culinary Chronicles* (Spring 2007), http://www.culinaryhistorians.ca/newsletters/CC_52.pdf. Further detail on Spadina can be found at www.toronto.ca/explore-enjoy/history-art-culture /museums/spadina-museum.

Chapter 8: GEORGIA O'KEEFFE KITCHEN

A most informative account is *Georgia O'Keeffe and Her Houses: Ghost Ranch and Abiquiu* by Barbara Buhler Lynes and Agapita Judy Lopez (New York: Abrams, in association with the Georgia O'Keeffe Museum, 2012). Chef Allen Smith of the Santa Fe School of Cooking gave the Georgia O'Keeffe cooking class along with former O'Keeffe cook Margaret Wood. Wood's books include *A Painter's Kitchen: Recipes from the Kitchen of Georgia O'Keeffe* (1997; reprint, Santa Fe: Museum of New Mexico Press, 2009) and *Remembering Miss O'Keeffe: Stories from Abiquiu* (Santa Fe: Museum of New Mexico Press, 2012).

Chapter 9: FRANK LLOYD WRIGHT KITCHEN

The experience of cooking in a Frank Lloyd Wright kitchen was made possible by the generosity of the Palumbo family, owners of Kentuck Knob, and Emily Butler, head of preservation and conservation at Kentuck Knob. Sources used in this chapter include *Kentuck Knob*, published by the Kentuck Knob House Museum, contact Kentuck Knob, 723 Kentuck Road, Chalk Hill, Pennsylvania 15421 or visit kentuckknob.com; *Frank Lloyd Wright: From within Outward*, by Richard Cleary et al. (New York:

Skira Rizzoli/Guggenheim Museum, 2009); *Architrecture of the Sun: Los Angeles Modernism 1900–1970* by Thomas S. Hines (Rizzoli, 2010); and *Pride of Place: Building the American Dream* by Robert A. M. Stern (Boston: Houghton Mifflin, 1986). Notes on mid-century modern architecture are from the book *Southern California in the '50s: Sun, Fun and Fantasy* by Charles Phoenix (Los Angeles: Angel City Press, 2001).

Chapter 10: JULIA CHILD KITCHEN

The history of Julia Child and her kitchen are largely taken from the exhibition and website "FOOD: Transforming the American Table 1950–2000," curated by Paula Johnson, at the Smithsonian Institution's National Museum of American History, http://americanhistory.si.edu/food/julia-childs-kitchen. The excellent Julia Child cheese soufflé cooking experience was given by David Lillico. Details on the life of Julia Child can be found in *Dearie: The Remarkable Life of Julia Child* by Bob Spitz (New York: A. A. Knopf, 2012). Information and quotes in this chapter are found in Julia Child's article "Julia Child at Home in Cambridge, Massachusetts," *Architectural Digest* (July/August 1976), https://www.architecturaldigest.com/story/julia-child-cambridge-massachusetts-house-article. Kathy Orton's article "Julia Child Slept Here," *Washington Post Magazine*, December 27, 2012, describes the Big Garland stove. An informative architectural article on the kitchen is "Julia's Kitchen: A Design Anatomy" by Bill Stumpf and Nicholas Polites, *Design Quarterly*, no. 104 (1977). Further information on the kitchen can be found in an article, "How to Arrange Your Kitchen According to Julia Child," August 17, 2016, by Pamela Heyne, https://lithub.com/how-to-arrange-your-kitchen-according-to-julia-child/.

Chapter 11: LOUIS ARMSTRONG KITCHEN

Louis Armstrong's life, house and love of eating were presented to me by David Reese, curator of the Louis Armstrong House Museum. Further information on the Armstrong kitchen can be found in an article by Stephen Patience, "Berth of the Blues," *The World of Interiors*, no. 344, and at www.stephenpatience.co.uk/Essays/Satchmo.html. Insights into the life of Lucille Armstrong can be found in "The Lucille Armstrong Story: A Lady with a

Vision" by Carolyn Carter-Kennedy, iUniverse, 2010. More details on the house can be found at www.louisarmstronghouse.org.

Chapter 12: ELVIS PRESLEY KITCHEN

Recipes can be found in *Fit for a King: The Elvis Presley Cookbook* by Elizabeth McKeon, Ralph Gevirtz and Julie Bandy (Nashville: Rutledge Hill Press, 1992). Douglas Martin is the author of the informative obituary "Mary Jenkins Langston, 78, Cook for Elvis Presley," *New York Times*, June 5, 2000. Further information on Graceland and its kitchen can be found at www.graceland.com and danielscottkitchens.co.uk/blog/a-look-inside-elvis-presleys-kitchen-in-graceland.

Chapter 13: PEARLSTONE KITCHEN

Information on building the house, Pacific Northwest food and cooking stir-fried spot prawns came from Joyce and Paul Pearlstone and Grace Kamitakahara. I was also aided by an article on the house and Vancouver Island life, "The Creative Synergy of Home and Garden" by Linda M. Langwith, *Seaside Magazine* (August 2013).